Matthias Haldimann

Fracture Strength of Structural Glass Elements

Matthias Haldimann

Fracture Strength of Structural Glass Elements

Analytical and Numerical Modelling, Testing and Design

Südwestdeutscher Verlag für Hochschulschriften

Impressum/Imprint (nur für Deutschland/ only for Germany)
Bibliografische Information der Deutschen Nationalbibliothek: Die Deutsche Nationalbibliothek verzeichnet diese Publikation in der Deutschen Nationalbibliografie; detaillierte bibliografische Daten sind im Internet über http://dnb.d-nb.de abrufbar.
Alle in diesem Buch genannten Marken und Produktnamen unterliegen warenzeichen-, marken- oder patentrechtlichem Schutz bzw. sind Warenzeichen oder eingetragene Warenzeichen der jeweiligen Inhaber. Die Wiedergabe von Marken, Produktnamen, Gebrauchsnamen, Handelsnamen, Warenbezeichnungen u.s.w. in diesem Werk berechtigt auch ohne besondere Kennzeichnung nicht zu der Annahme, dass solche Namen im Sinne der Warenzeichen- und Markenschutzgesetzgebung als frei zu betrachten wären und daher von jedermann benutzt werden dürften.

Verlag: Südwestdeutscher Verlag für Hochschulschriften Aktiengesellschaft & Co. KG
Dudweiler Landstr. 99, 66123 Saarbrücken, Deutschland
Telefon +49 681 37 20 271-1, Telefax +49 681 37 20 271-0, Email: info@svh-verlag.de
Zugl.: Lausanne, EPFL, Diss., 2006

Herstellung in Deutschland:
Schaltungsdienst Lange o.H.G., Berlin
Books on Demand GmbH, Norderstedt
Reha GmbH, Saarbrücken
Amazon Distribution GmbH, Leipzig
ISBN: 978-3-8381-0534-5

Imprint (only for USA, GB)
Bibliographic information published by the Deutsche Nationalbibliothek: The Deutsche Nationalbibliothek lists this publication in the Deutsche Nationalbibliografie; detailed bibliographic data are available in the Internet at http://dnb.d-nb.de.
Any brand names and product names mentioned in this book are subject to trademark, brand or patent protection and are trademarks or registered trademarks of their respective holders. The use of brand names, product names, common names, trade names, product descriptions etc. even without a particular marking in this works is in no way to be construed to mean that such names may be regarded as unrestricted in respect of trademark and brand protection legislation and could thus be used by anyone.

Publisher:
Südwestdeutscher Verlag für Hochschulschriften Aktiengesellschaft & Co. KG
Dudweiler Landstr. 99, 66123 Saarbrücken, Germany
Phone +49 681 37 20 271-1, Fax +49 681 37 20 271-0, Email: info@svh-verlag.de

Copyright © 2009 by the author and Südwestdeutscher Verlag für Hochschulschriften Aktiengesellschaft & Co. KG and licensors
All rights reserved. Saarbrücken 2009

Printed in the U.S.A.
Printed in the U.K. by (see last page)
ISBN: 978-3-8381-0534-5

Abstract

For centuries, the use of glass in buildings was essentially restricted to functions such as windows and glazing. Over the last decades, continuous improvements in production and refining technologies have enabled glass elements to carry more substantial superimposed loads and therefore achieve a more structural role. The structural design of such elements, however, remains problematic.

Current widely used design methods suffer from notable shortcomings. They are, for instance, not applicable to general conditions, but are limited to special cases like rectangular plates, uniform lateral loads, constant loads, time-independent stress distributions and the like. Some model parameters combine several physical aspects, so that they depend on the experimental setup used for their determination. The condition of the glass surface is not represented by user-modifiable parameters, but is embedded implicitly. The design methods contain inconsistencies and give unrealistic results for special cases. Different models yield differing results and several researchers have expressed fundamental doubts about the suitability and correctness of common glass design methods. The lack of confidence in 'advanced' glass models and the absence of a generally agreed design method result in frequent time-consuming and expensive laboratory testing and in inadequately designed structural glass elements. The present thesis endeavours to improve this situation.

After outlining the fundamental aspects of the use of glass as a building material, an analysis of present knowledge was conducted in order to provide a focus for subsequent investigations. Then a lifetime prediction model for structural glass elements was established based on fracture mechanics and the theory of probability. Aiming at consistency, flexibility, and a wide field of application, this model offers significant advantages over currently used models. It contains, for instance, no simplifying hypotheses that would restrict its applicability to special cases and it offers great flexibility with regard to the representation of the surface condition. In a next step, possible simplifications of the model and the availability of the model's required input data were discussed. In addition to the analysis of existing data, laboratory tests were performed and testing procedures improved in order to provide more reliable and accurate model input.

In the last part of the work, recommendations for structural design and testing were developed. They include, among other things, the following:

- Glass elements that are permanently protected from damage can be designed by extrapolation of experimental data obtained from as-received or homogeneously damaged specimens. The design of exposed glass elements whose surfaces may be damaged during their service lives (for example, because of accidental impact or vandalism), however, should be based on a realistic estimation of the potential damage (design flaw). Appropriate predictive models and testing procedures are proposed in this thesis.
- If substantial surface damage has to be considered, the inherent strength contributes little to the resistance of heat treated glass. Therefore, quality control measures that allow the use of a high design value for the residual surface stress are very efficient in terms of economical material use.
- Results from laboratory testing at ambient conditions represent a combination of surface condition and crack growth. The strong stress rate dependence of the latter, which was demonstrated in this thesis, diminishes the accuracy and reliability of the results. The problem can be addressed by the near-inert testing procedure that was developed and used in this thesis.

The application of the proposed models and recommendations in research and practice is facilitated by *GlassTools*, the computer software that was developed as part of this thesis.

Keywords: *Design method, fracture mechanics, fracture strength, laboratory testing procedure, predictive modelling, structural design, structural glass elements, surface flaws, resistance.*

Zusammenfassung

Während Jahrhunderten wurde Glas in Gebäuden praktisch ausschliesslich für Fenster und Verglasungen eingesetzt. In den letzten Jahrzehnten haben Fortschritte in der Herstellung und Verarbeitung von Glas den zunehmenden Einsatz von Glasbauteilen für lastabtragende Elemente ermöglicht. Die Bemessung solcher Bauteile bleibt jedoch problematisch.

Die gängigen Bemessungsmethoden weisen wesentliche Mängel auf. Sie sind beispielsweise nicht allgemein anwendbar, sondern auf Spezialfälle wie rechteckige Glasscheiben, gleichmässig verteilte Flächenlasten, konstante Lasten, zeitunabhängige Spannungsverteilungen und dergleichen beschränkt. Einige Modellparameter kombinieren mehrere physikalische Aspekte und hängen daher von der zu ihrer Bestimmung verwendeten Versuchseinrichtung ab. Der Zustand der Glasoberfläche wird nicht durch vom Anwender anpassbare Parameter berücksichtigt, sondern ist implizit im Modell enthalten. Darüber hinaus sind die Bemessungskonzepte nicht durchgängig konsistent, sie führen zu unrealistischen Resultaten für Spezialfälle und die Ergebnisse der Modelle unterscheiden sich deutlich. Verschiedene Forscher äusserten sogar grundlegende Zweifel an der Zweckmässigkeit und Richtigkeit gängiger Bemessungsmethoden. Die weitverbreitete Skepsis gegenüber 'fortgeschrittenen' Modellen und das Fehlen einer allgemein anerkannten Bemessungsmethode führen zu häufigen zeit- und kostenintensiven Laborversuchen und zu unzulänglich bemessenen Glasbauteilen. Diese Situation zu verbessern ist das Ziel der vorliegenden Dissertation.

Nach einem kurzen Überblick über die grundlegenden Aspekte des Einsatzes von Glas als Baustoff wird der aktuelle Wissensstand analysiert, um die nachfolgenden Untersuchungen zu fokussieren. Anschliessend wird auf der Basis der linear elastischen Bruchmechanik und der Wahrscheinlichkeitstheorie ein möglichst allgemeines, konsistentes und flexibles Modell zur Vorhersage der Lebensdauer von Bauteilen aus Glas entwickelt. Dieses bietet gegenüber gängigen Modellen wesentliche Vorteile. Es enthält beispielsweise keine den Anwendungsbereich einschränkenden Annahmen und bietet grosse Flexibilität bei der Modellierung des Oberflächenzustandes. Möglichkeiten zur Vereinfachung des Modells sowie die Verfügbarkeit der benötigten Eingabedaten werden diskutiert. Neben der Analyse publizierter Daten werden dazu auch eigene Laborversuche durchgeführt. Schliesslich werden bestehende Prüfverfahren in Bezug auf die Zuverlässigkeit und Genauigkeit der Ergebnisse optimiert.

Im letzten Teil der vorliegenden Arbeit werden Empfehlungen für die Bemessung sowie für die Durchführung von Laborversuchen erarbeitet. Es wird unter anderem Folgendes festgestellt:

♦ Für dauerhaft vor Beschädigung geschützte Glasbauteile liefert eine Bemessung durch Extrapolation der Ergebnisse von Versuchen an neuen oder homogen vorgeschädigten Gläsern sinnvolle Ergebnisse. Exponierte Elemente, deren Oberfläche während der Lebensdauer z. B. durch Anprall oder Vandalismus beschädigt werden könnte, sollten jedoch anhand einer realistischen Abschätzung der potentiellen Beschädigung (Bemessungsriss) bemessen werden. Entsprechende Modelle und Laborversuche werden vorgeschlagen.

♦ Wenn wesentliche Oberflächenschädigungen zu berücksichtigen sind, trägt die Glaseigenfestigkeit nur wenig zum Widerstand von thermisch vorgespannten Gläsern bei. Qualitätssicherungsmassnahmen, welche die Verwendung eines hohen Bemessungswertes der Druckeigenspannung ermöglichen, sind daher sehr effektiv im Hinblick auf eine optimale Materialausnutzung.

♦ Ergebnisse aus Laborversuchen in normaler Umgebung werden sowohl durch den Oberflächenzustand als auch durch das unterkritische Risswachstum beeinflusst. Die Tatsache, dass das Risswachstum in der vorliegenden Dissertation gezeigt stark von der Belastungsgeschwindigkeit abhängt, reduziert die Genauigkeit und Sicherheit der Ergebnisse. Das Problem wird mit dem in der vorliegenden Arbeit entwickelten und angewendeten quasi-inerten Prüfverfahren vermieden.

Die Anwendung der Modelle und Empfehlungen in Forschung und Praxis wird duch *GlassTools*, eine im Rahmen dieser Arbeit entwickelte Software, wesentlich vereinfacht und beschleunigt.

Stichworte: *Bemessungsmethode, Bruchmechanik, Bruchfestigkeit, Prüfverfahren, Modellierung, Bemessung, tragende Glasbauteile, Oberflächenschädigung, Tragwiderstand.*

Résumé

Depuis des siècles, l'utilisation du verre dans les bâtiments était limitée essentiellement à ses fonctions de fenêtres et vitrages. Au cours des dernières décennies, des progrès constants, réalisés aussi bien dans la production que dans les techniques de traitement du verre, ont permis de développer des éléments en verre capables de supporter des charges utiles importantes, leur permettant ainsi de jouer un rôle structural. Cependant, le dimensionnement de tels éléments porteurs reste encore problématique.

Les méthodes de dimensionnement usuelles souffrent d'insuffisances considérables. Elles ne sont par exemple pas applicables aux cas généraux, mais sont limitées aux cas particuliers comme les panneaux rectangulaires, les charges transversales uniformément réparties, les forces constantes, les répartitions de contraintes indépendantes du temps, etc. Certains paramètres combinent plusieurs phénomènes physiques, si bien qu'ils dépendent du schéma expérimental qui a servi à leur détermination. L'état de surface du verre n'est pas modélisé par des paramètres qui peuvent être définis par l'utilisateur, mais est pris en considération implicitement. Les méthodes de dimensionnement contiennent des contradictions et peuvent donner des résultats non réalistes dans des cas particuliers. Des modèles différents conduisent à des résultats différents pour la même application. Enfin plusieurs chercheurs ont exprimé de sérieux doutes quant à l'applicabilité et la justesse des méthodes usuelles de calcul des éléments en verre. Le manque de confiance dans des modèles plus 'sophistiqués' ainsi que l'absence de méthodes universellement approuvées conduisent à la réalisation d'essais en laboratoire longs et coûteux et à des dimensionnements inadéquats d'éléments structuraux en verre. La présente thèse a pour objectif d'améliorer cette situation.

Après avoir rappelé les aspects fondamentaux de l'utilisation du verre en tant que matériau de construction, l'auteur a procédé à une analyse des connaissances actuelles de façon à définir l'orientation des recherches à poursuivre. Ensuite, un modèle de prévision de la durée de vie des éléments en verre, le plus cohérent, général et souple possible, a été établi sur la base de la mécanique de la rupture et de la théorie des probabilités. Ce modèle offre de nombreux avantages par rapport aux modèles existants: il ne contient par exemple aucune hypothèse simplificatrice qui limiterait son application à des cas spéciaux et il offre une grande souplesse dans la représentation de l'état de surface. L'étape suivante a consisté à discuter des éventuelles simplifications du modèle et de la disponibilité des données nécessaires à introduire dans le modèle. Enfin, en plus de l'analyse des résultats expérimentaux existants, des essais en laboratoire ont été effectués et des modes opératoires ont été améliorés de façon à fournir au modèle des données plus fiables et plus précises.

Dans la dernière partie de la thèse, des recommandations ont été proposées pour le dimensionnement structural et la réalisation des essais. Ces recommandations comprennent entre autres choses les points suivants:

- Les éléments en verre qui sont protégés en permanence vis-à-vis d'endommagements de surface peuvent être dimensionnés à partir de résultats d'essai obtenus sur des éprouvettes neuves ou endommagées uniformément. Le dimensionnement d'éléments en verre exposés, dont la surface peut subir des endommagements pendant sa durée de service (chocs, vandalisme), doit par contre être basé sur une estimation réaliste du dommage potentiel (défaut de calcul). Des modèles de prévision appropriés et des modes opératoires sont proposés dans la thèse.
- Si un dommage de surface important doit être pris en considération lors du dimensionnement d'un verre trempé thermiquement, la résistance intrinsèque du verre participe peu à la résistance. C'est pourquoi des mesures de contrôle de qualité, qui permettent de tenir compte d'une valeur élevée des contraintes résiduelles, sont très efficaces en vue d'une utilisation économique de matériau.
- Les résultats d'essai en laboratoire sous conditions ambiantes représentent une combinaison de l'état de surface et de l'accroissement des fissures. La forte dépendance de cet accroissement à la vitesse de contrainte, démontrée dans ce projet de recherche, diminue la précision et la fiabilité

des résultats. Le problème peut être résolu à l'aide d'essais réalisés selon le mode opératoire 'quasi-inerte' développé et utilisé dans cette thèse.

Le logiciel GlassTools, développé par l'auteur dans le cadre de cette thèse, facilite l'application, aussi bien en recherche qu'en pratique, des modèles et recommandations proposés.

Mots-clés: *Méthode de dimensionnement, mécanique de la rupture, résistance à la rupture, mode opératoire, modèle de prévision, dimensionnement, éléments structuraux en verre, anomalie de surface, résistance*

Acknowledgements

First, I would like to thank Prof. Dr. Manfred A. Hirt, director of the Steel Structures Laboratory ICOM and of this thesis, as well as Michel Crisinel, the supervisor of this thesis, for giving me the opportunity to work in an excellent research environment, for their support, and for their feedback on my work, which improved its quality significantly.

I would like to thank COST Switzerland (Swiss State Secretariat for Education and Research SER) and the Swiss National Science Foundation SNF for partially funding this thesis. Thanks are furthermore due to Dr. Roman Graf, Heinz Misteli, and Rudolf Bernhard from Glas Trösch AG (Bützberg) for donating the glass specimens for the experiments and sharing their expertise.

Numerous people were involved in the guidance and assessment of this thesis. I would like to offer my thanks to the members of the advisory committee, who met periodically over the course of the thesis to offer advice, ideas, and suggestions: Prof. Dr. Robert Dalang (EPFL), Prof. Dr. Michael H. Faber (ETH Zurich), Dr. Roman Graf (Glas Trösch AG, Bützberg), Dr. Rudolf Hess (Glasconsult, Uitikon), and Edgar Joffré (Félix Constructions SA, Bussigny-Lausanne). The advisory committee meetings were highly valued and had a significant influence on the direction of the work. I would also like to sincerely thank the members of the defence jury, which consisted of Prof. Dr. Geralt Siebert (Bundeswehr University Munich, Germany), Prof. Dr. Michael H. Faber and Dr. Rudolf Hess and was chaired by Prof. Dr. Anton Schleiss (EPFL).

Over the course of the thesis, I have come into contact with a number of other experts, whose advice I have greatly appreciated. My very special thanks go to Dr. Mauro Overend (University of Nottingham, United Kingdom), for taking the time to read the first draft of the thesis document and providing me with invaluable feedback that has significantly elevated the quality of the final document. My very special thanks go also to Dr. Andreas Luible (Josef Gartner Switzerland AG), my friend and predecessor in the research area of structural glass at ICOM, for generously sharing his knowledge and experience. In this limited space, I would furthermore like to thank specifically for their advice: Christoph Haas (Ernst Basler + Partner AG, Zurich), Dr. Wilfried Laufs (WernerSobekNewYork Inc., USA), Dr. Jørgen Munch-Andersen (Danish Building Research Institute), Dr. Alain Nussbaumer (EPFL), Dr. Jens Schneider (Goldschmidt Fischer und Partner GmbH, Germany) and all members of the European COST Action C13 'Glass & Interactive Building Envelopes'. Thanks are also due to Steve Keating, who mastered the difficult job of proofreading this highly specialized scientific text.

The experimental work conducted was made possible by the competent staff of the structural laboratory, namely Sylvain Demierre, Roland Gysler, Gilbert Pidoux, and Hansjakob Reist, as well as by the two student assistants Giovanni Accardo and Hugues Challes, who were a great and loyal help. My thanks also to Danièle Laub (EPFL) for providing microscopy equipment and knowledge. My thanks also to Esther von Arx and Claudio Leonardi for their much-valued help with administrative and organizational issues.

I much enjoyed the exceptionally pleasant atmosphere at work and outside work, as well as the good cooperation on various non-thesis related projects. In addition to the past and present ICOM members already mentioned, I would therefore like to thank Abeer, Ahti, Ann, Bertram, Claire, Danijel, François, Jean-Paul, Laurence, Luis, Michel T., Philippe, Pierre, Rahel, Scott, Stéphane, Tamar, Thierry, Thomas, and Yves.

I would like to express my deep gratitude to my family for their interest in my work and their support, which were of inestimable value during my long academic education. My parents, my parents-in-law and my sister Mirjam receive my most sincere thanks for the many times they took excellent care of our daughter Vera.

Finally and most importantly, I am immensely grateful and deeply indebted to my wife, Senta, for her unconditional support and her encouragement, for the countless interesting discussions on the content of this thesis, for her understanding and patience when I had to work evenings and week-ends, and for the wonderful time that we have spent since we met at ICOM shortly before I began this thesis.

Contents

Abstract	i
Zusammenfassung	ii
Résumé	iii
Acknowledgements	v
Contents	vii
Notation, Definitions, Terminology	**xi**
General information	xi
Latin symbols	xi
Greek symbols	xiii
Generally used indices and superscripts	xiv
Functions and mathematical notation	xiv
Abbreviations	xiv
Glossary of Terms	**xv**
1 Introduction	**1**
1.1 Background and motivation	1
1.2 Main objectives	3
1.3 Organization of the thesis	3
2 Glass for Use in Buildings	**5**
2.1 Production of flat glass	5
2.2 Material properties	6
2.2.1 Composition and chemical properties	6
2.2.2 Physical properties	8
2.3 Processing and glass products	9
2.3.1 Introduction	9
2.3.2 Tempering of glass	10
2.3.3 Laminated glass	13

3 State of Knowledge – Overview and Analysis   15

3.1 Laboratory testing procedures   15
3.1.1 Static fatigue tests   15
3.1.2 Dynamic fatigue tests   15
3.1.3 Direct measurement of the growth of large through-thickness cracks   17
3.1.4 Tests for the glass failure prediction model   17

3.2 Design methods   17
3.2.1 Introduction   17
3.2.2 Allowable stress based methods   17
3.2.3 DELR design method   18
3.2.4 European draft standard prEN 13474   20
3.2.5 Design method of Shen   22
3.2.6 Design method of Siebert   23
3.2.7 Glass failure prediction model (GFPM)   25
3.2.8 American National Standard ASTM E 1300   25
3.2.9 Canadian National Standard CAN/CGSB 12.20   27

3.3 Analysis of current glass design methods   29
3.3.1 Introduction   29
3.3.2 Main concepts   29
3.3.3 Obtaining design parameters from experiments   29
3.3.4 Load duration effects   30
3.3.5 Residual stress   32
3.3.6 Size effect   32
3.3.7 Weibull statistics   33
3.3.8 Biaxial stress fields   33
3.3.9 Consistency and field of application   34
3.3.10 Additional information and quality control   34

3.4 Summary and Conclusions   35

4 Lifetime Prediction Model   37

4.1 Stress corrosion and subcritical crack growth   37
4.1.1 Introduction and terminology   37
4.1.2 The classic stress corrosion theory   37
4.1.3 Relationship between crack velocity and stress intensity   38
4.1.4 Chemical background   40
4.1.5 Crack healing, crack growth threshold and hysteresis effect   40
4.1.6 Influences on the relationship between stress intensity and crack growth   41

4.2 Mechanical behaviour of a single surface flaw   43
4.2.1 Fracture mechanics basics   43
4.2.2 Inert strength   44
4.2.3 Heat treated glass   45
4.2.4 Subcritical crack growth and lifetime   45
4.2.5 Equivalent static stress and resistance   46
4.2.6 Constant stress and constant stress rate   48

4.3 Extension to a random surface flaw population   49
4.3.1 Starting point and hypotheses   49
4.3.2 Constant, uniform, uniaxial stress   49
4.3.3 Extension to non-uniform, biaxial stress fields   52
4.3.4 Extension to time-dependent loading   54
4.3.5 Extension to account for subcritical crack growth   54

4.4 Summary and Conclusions   56

5 Discussion and Simplification of the Model — 57

5.1 Assessment of the main hypotheses behind the model — 57
- 5.1.1 Introduction — 57
- 5.1.2 Strength data — 58
- 5.1.3 In-service conditions — 60
- 5.1.4 Conclusions — 61

5.2 Assessment of the relevance of some fundamental aspects — 62
- 5.2.1 The risk integral — 62
- 5.2.2 The crack growth threshold — 64
- 5.2.3 Biaxial stress fields — 66

5.3 Simplification for special cases — 71
- 5.3.1 Structural Design — 71
- 5.3.2 Avoiding transient finite element analyses — 72
- 5.3.3 Common testing procedures — 75

5.4 Relating current design methods to the lifetime prediction model — 76
5.5 Summary and Conclusions — 78

6 Quantification of the Model Parameters — 79

6.1 Introduction — 79
6.2 Basic fracture mechanics parameters — 80
6.3 Crack velocity parameters — 82
6.4 Ambient strength data — 86
6.5 Analysing large-scale experiments — 88
6.6 Residual surface stress data — 92
6.7 Summary and Conclusions — 95

7 Experimental Investigations — 97

7.1 Objectives — 97
7.2 Test setup and measurements — 98
7.3 Creating near-inert conditions in laboratory tests — 104
7.4 Deep close-to-reality surface flaws — 105
- 7.4.1 Testing procedure — 105
- 7.4.2 Experiments and results — 108
- 7.4.3 Is hermetic coating indispensable for deep flaw testing? — 110

7.5 Visual detectability of deep surface flaws — 111
7.6 Summary and Conclusions — 114

8 Structural Design of Glass Elements — 115

8.1 Introduction — 115
8.2 The structural design process — 116
8.3 Characteristics of the two surface condition models — 116
- 8.3.1 Single surface flaw model — 116
- 8.3.2 Random surface flaw population model — 118

8.4 Recommendations — 123
- 8.4.1 Structural design — 123
- 8.4.2 Testing — 125
- 8.4.3 Overview of mathematical relationships — 126

8.5 Software — 128
- 8.5.1 Introduction and scope — 128
- 8.5.2 Discretization of the failure prediction model — 128

	8.5.3 Implementation	130
	8.5.4 Features and usage notes	132
8.6	Action modelling	133
8.7	Summary and Conclusions	134

9 Summary, Conclusions, and Further Research 137
9.1 Summary 137
9.2 How this work's objectives were reached 137
9.3 Main conclusions 138
9.4 Further research 141

References 143

A Overview of European Standards 153

B On the Derivation of the Lifetime Prediction Model 155

C EPFL-ICOM Laboratory Testing Information and Data 157

D Ontario Research Foundation Experimental Data 161

E Statistical Fundamentals 165

F Predictive Modelling Example 169

G Source Code Listings 171

H Documentation for the 'GlassTools' Software 179

Index 197

Curriculum vitae 201

Notation, Definitions, Terminology

General information

Variables are defined and explained on their first occurrence only. In case of doubt, readers should refer to the symbol list below. It gives a short description of all variables as well as references to the place where they are defined in the text.

Particularly unfamiliar or important terms are defined in the glossary (p. xv). On their first occurrence in the text, these terms are marked with a small upwards arrow[↑]. Additionally, there is an extensive index at the very end of the document.

The present document follows current regulations on technical and scientific typesetting, in particular ISO 31-0:1992, ISO 31-0:1992/Amd.2:2005, ISO 31-3:1992 and ISO 31-11:1992. Accordingly, *italic* symbols are used only to denote those entities that may assume different values. These are typically physical or mathematical variables. Symbols, including subscripts and superscripts, which do not represent physical quantities or mathematical variables are set in upright roman characters. (Example: The exponent 'n' (italic) in σ_n^n is a physical variable, while the index 'n' (roman) is an abbreviation for 'normal'.)

Latin symbols

a	1. crack depth → p. 45; 2. long edge length of a rectangular plate → p. 25
a_0	lower limit of the crack depth → p. 50
a_i	initial crack depth → p. 45
a_c	critical crack depth → p. 44
a_f	crack depth at failure → p. 46
a_th	threshold crack depth → p. 65
a_VDL	lower visual detectability limit → p. 112
b	short edge length of a rectangular plate → p. 25
c	dimensionless stress distribution function → p. 72
f_0	reference ambient strength → p. 72
$f_{0,\mathrm{inert}}$	reference inert strength → p. 51
h	effective glass thickness → p. 25

i, j, k	integer variables
$\bar{k}$	combined parameter (used to simplify notation); → p. 72
$\tilde{k}$	first surface flaw parameter of the glass failure prediction model → p. 25
m_0	second surface condition parameter (see also θ_0) → p. 51
$\bar{m}$	combined parameter (used to simplify notation) → p. 67
$\tilde{m}$	second surface flaw parameter of the glass failure prediction model → p. 25
n	exponential crack velocity parameter → p. 39
p_{AD}	observed significance level probability (p-value) of the Anderson-Darling goodness-of-fit test → p. 168
q	pressure, uniformly distributed load
q^*	normalized pressure → p. 68
$\tilde{q}$	non-dimensionalized load → p. 25
r	parameter of the PDF of the crack depth → p. 50
$\vec{r}$	a point on a surface (defined by two coordinates x and y, $\vec{r} = \vec{r}(x, y)$) → p. 52
t	time
t_0	reference time period → p. 46
t_f	1. time to failure; 2. point in time when failure occurs; 3. lifetime
$t_{f,n}$	normalized lifetime → p. 47
t_n	normalized loading time → p. 47
$t_{res,n}$	normalized residual lifetime → p. 48
v	crack velocity → p. 38
v_0	linear crack velocity parameter (reference crack velocity) → p. 39
x, y	coordinates of a point on a surface, cf. $\vec{r}$
A	surface area (general)
A_0	unit surface area ($= 1\,\mathrm{m}^2$) → p. 51
A_{red}	decompressed surface area → p. 19
B	Weibull's risk function → p. 25
D	plate flexural rigidity → p. 68
D_{res}	relative residual lifetime → p. 48
E	Young's modulus → p. 44;
$\hat{F}$	empirical cumulative distribution function → p. 166
I, J, K	the total number of the quantity counted with i, j, k ($I = \max(i), J = \max(j), K = \max(k)$)
K_I	stress intensity factor for fracture mode I loading (opening mode) → p. 43
K_{Ic}	fracture toughness (critical stress intensity factor) for fracture mode I loading → p. 43
K_{th}	threshold stress intensity factor → p. 40
L	likelihood function → p. 167
N	number of samples or other countable quantity
P_f	failure probability ($= 1 - P_s$) → p. 49
$P_{f,t}$	target failure probability → p. 72
P_s	survival probability ($= 1 - P_f$) → p. 50
Q	force, point load, location- and orientation dependent failure probability
R	principal stress ratio ($R = \sigma_2/\sigma_1$) → p. 67; resistance
S	summed square of residuals → p. 167
T	duration, time period or end of a time period starting at $t = 0$
U	coefficient combining fracture mechanics and crack velocity parameters → p. 55
Y	geometry factor (caution: does *not* include $\sqrt{\pi}$; it is $K_I = Y\sqrt{\pi} \cdot \sigma_n \cdot \sqrt{a}$) → p. 43

Greek symbols

β	1. shape parameter of the Weibull distribution → p. 17; 2. reliability index → p. 119
γ	partial factor (specified more precisely in the index)
η_b	biaxial stress correction factor → p. 67
θ	general Weibull scale parameter (specified more precisely in the index)
θ_0	first surface condition parameter (see also m_0) → p. 51
μ	mean
ν	Poisson's ratio → p. 44
σ	1. crack opening surface stress → p. 43 (unless otherwise stated, compressive stresses are negative and tensile stresses positive); 2. standard deviation
$\dot{\sigma}$	stress rate ($\dot{\sigma} = d\sigma/dt$)
$\bar{\sigma}$	equivalent reference stress → p. 71
$\check{\sigma}$	representative stress (often $\sigma_{\max}$) → p. 72
$\tilde{\sigma}$	non-dimensionalized stress → p. 25
σ_{01}	unit stress = 1 MPa (for normalization) → p. 47
σ_1	major in-plane principal stress ($\sigma_1 \geq \sigma_2$) → p. 53
σ_2	minor in-plane principal stress ($\sigma_1 \geq \sigma_2$) → p. 53
σ_c	inert strength of a crack (also called 'critical stress') → p. 44
σ_{t_0}	t_0-equivalent static stress → p. 46
σ_E	surface stress due to action(s) E → p. 45
σ_f	stress at failure (also known as 'failure stress' or 'breakage stress')
$\sigma_{\max}$	maximum principal stress in an element (geometric maximum) → p. 72
σ_n	stress normal to a crack's plane → p. 53
σ_r	residual surface stress due to tempering (sometimes called 'prestress'; compressive ⇒ negative sign) → p. 45
σ_{R,t_0}	t_0-equivalent resistance → p. 47
σ_p	stress due to external constraints or prestressing → p. 45
φ	crack orientation, angle → p. 53
$\hat{\varphi}$	upper integral boundary for η_b → p. 67
τ	1. time (point in time); 2. shear stress (general) → p. 53
ψ	load combination factor → p. 27
Λ	logarithmic likelihood function → p. 167

Generally used indices and superscripts

$X_{I, II, III}$	related to crack mode I, II or III	X_{inert}	in or for inert conditions
X_{adm}	admissible	X_i	i-th value, case or time period
X_c	critical	X_n	normal, normalized, national
X_d	design level	X_{test}	in laboratory testing, in laboratory conditions
X_E	due to actions		
X_{eff}	effective	$\sigma^{(i)}$	i-th value, case or time period (avoids σ_1 and σ_2, which are the principal stresses)
X_{eq}	equivalent		
X_f	failure, at failure, related to failure	$x^{(1)}$	related to a single crack
X_i	initial	$x^{(k)}$	related to k cracks

Functions and mathematical notation

X	placeholder for any variable or text	$\ll$	much less
$[X]$	the unit of X	$\parallel$	parallel to
$\|X\|$	the absolute value of X	$f(X)$	a function of the variable X
$\langle X \rangle$	a parameter of a function (to be replaced by the actual value of what is described in X)	$\Gamma()$	the Gamma function
		$\max()$	maximum
X^T	the matrix transpose of X (the matrix obtained by exchanging X's rows and columns)	$\min()$	minimum
		$\ln()$	natural logarithm
$\forall$	for all (also 'for each' or 'for every')	$\mathscr{P}(X)$	probability of the event X ($0 \leq \mathscr{P}(X) \leq 1$)
$\exists$	there exists		
$\propto$	proportional to	Φ	the cumulative distribution function of the standard normal distribution
$\gg$	much greater		

Abbreviations

4PB	four point bending (test setup)	HSG	heat strengthened glass
ANG	annealed glass	IPP	in-plane principal stress
BSG	borosilicate glass	LEFM	linear elastic fracture mechanics
CDF	cumulative distribution function	LR	loading ring (in a CDR test setup)
CDR	coaxial double ring (test setup)	ORF	Ontario Research Foundation
CS	constant stress	PDF	probability density function
CSR	constant stress rate	PVB	polyvinyl butyral
DCB	double cantilever beam (test type)	RSFP	random surface flaw population
DT	double torsion (test type)	SCG	subcritical crack growth
FE	finite element	SIF	stress intensity factor
FTG	fully tempered glass	SLS	soda lime silica glass
GFPM	glass failure prediction model	SSF	single surface flaw

Glossary of Terms

Action General term for all mechanical, physical, chemical and biological actions on a structure or a structural element, e. g. pressures, loads, forces, imposed displacements, constraints, temperature, humidity, chemical substances, bacteria and insects.

Action history The description of an action as a function of time.

Action intensity The magnitude of an action, e. g. a load intensity, a stress intensity or the magnitude of an imposed deformation. See also 'load shape'.

Addend A number (or a mathematical expression that evaluates to a number) which is to be added to another.

Air side In the float process, the upper side of glass is called the air side.

Alkali Substance that neutralizes acid to form salt and water. Yields hydroxyl (OH-) ions in water solution. Proton acceptor.

Ambient temperature The environmental temperature surrounding the object.

Annealing The process which prevents glass from shattering after it has been formed. The outer surfaces of the glass shrink faster than the glass between the surfaces, causing residual stresses which can lead to shattering. This can be avoided by reheating the glass and allowing it to cool slowly.

Artificially induced surface damage Any kind of damage that is induced systematically and on purpose, e. g. for laboratory testing. If it is homogeneous in terms of its characteristics and its distribution on the surface, it is called 'artificially induced homogeneous surface damage' (e. g. surface damage induced by sandblasting).

As-received glass Glass as it is delivered to the client, sometimes also called 'new glass'. The surface contains only the small and random flaws introduced by production, cutting, handling and shipping.

Aspect ratio The relationship between the long and the short edge lengths of a rectangular plate.

Autoclave A vessel that employs heat and pressure. In the glass industry, used to produce a bond between glass and PVB or urethane sheet, thus creating a laminated sheet product.

Bevelling The process of edge-finishing flat glass by forming a sloping angle to eliminate right-angled edges.

Coating A material, usually liquid, used to form a covering film over a surface. Its function is to decorate, to protect the surface from destructive agents or environments (abrasion, chemical action, solvents, corrosion, and weathering) and/or to enhance the (optical, mechanical, thermal) performance.

Coefficient of variation (CoV) A measure of dispersion of a probability distribution. It is defined as the ratio of the standard deviation σ to the mean μ.

Computing time The time required to run an algorithm on a computer. While the actual value depends on the performance on the computer, the term is still useful for qualitative considerations and comparisons.

Constant stress rate loading A specimen is loaded such that the stress increases linearly with respect to time.

Constant load rate loading The load on a specimen is increased linearly with respect to time.

Convection A transfer of heat through a liquid or gas, when that medium hits against a solid.

Crack In the present document, the term 'crack' refers to the idealized model of a flaw having a defined geometry and lying in a plane.

Cullet Recycled or waste glass.

Curtain walling Non-load bearing, typically aluminium, façade cladding system, forming an integral part of a building's envelope.

Decompressed surface The part of an element's surface where the tensile stress due to loading is greater that the residual compressive stress due to tempering. On these parts of the surface, there is a positive crack opening stress.

Defect A flaw that is unacceptable.

Deflection The physical displacement of glass from its original position under load.

Deformation Any change of form or shape produced in a body by stress or force.

Desiccants Porous crystalline substances used to absorb moisture and solvent vapours from the air space of insulating glass units. More properly called absorbents.

Design See structural design.

Design life The period of time during which a structural element is expected to perform according to its specification, i. e. to meet the performance requirements.

Double glazing, double-glazed units See insulating glass unit.

Effective nominal flaw depth The depth of a flaw that is calculated from its measured strength.

Elasticity The property of matter which causes it to return to its original shape after deformation, such as stretching, compression, or torsion.

Elongation Increase in length expressed numerically as a fraction or percentage of initial length.

Emissivity The relative ability of a surface to absorb and emit energy in the form of radiation. Emissivity factors range from 0.0 (0%) to 1.0 (100%).

Equibiaxial stress field The two in-plane principal stresses are equal ($\sigma_1 = \sigma_2$). In this stress state, the stress normal to a crack σ_n is independent of the crack's orientation φ_{crack}, meaning that $\sigma_n = \sigma_1 = \sigma_2\ \forall \varphi_{crack}$. An equibiaxial stress field is in particular found within the loading ring in coaxial double ring testing.

Face Describes the surfaces of the glass in numerical order from the exterior to the interior. The exterior surface is always referred to as face 1. For a double-glazed unit, the surface of the outer pane facing into the cavity is face 2, the surface of the inner pane facing into the cavity is face 3 and the internal surface of the inner pane is face 4.

Flat glass Pertains to all glass produced in a flat form.

Flaw General term describing a condition or change that indicates an abnormal condition or imperfection in a material. Only flaws that are unacceptable are defects.

Float glass Transparent glass with flat, parallel surfaces formed on the surface of a bath of molten tin. If no information with respect to heat treatment is given, the term generally refers to *annealed* float glass.

Fully tempered glass Glass with a high residual compressive surface stress, varying typically between 80 MPa and 150 MPa in the case of soda lime silica glass. According to ASTM C 1048-04, fully tempered glass is required to have either a minimum surface compression of 69 MPa (10 000 psi) or an edge compression of not less than 67 MPa (9 700 psi). In European standards, the fragmentation count and the maximum fragment size is specified (EN 12150-1:2000; EN 12150-2:2004).

Glass A uniform amorphous solid material, usually produced when a suitably viscous molten material cools very rapidly to below its glass transition temperature, thereby not giving enough time for a regular crystal lattice to form. By far the most familiar form of glass is soda lime silica glass. In its pure form, glass is a transparent, relatively strong, hard-wearing, essentially inert, and biologically inactive material which can be formed with very smooth and impervious surfaces. Glass is, however, brittle and will break into sharp shards. These properties can be modified, or even changed entirely, through the addition of other compounds or heat treatment.

Glazing The securing of glass into prepared openings. It also refers to the collective elements of a building comprising glass, frame and fixings.

Hardness Property or extent of being hard. Measured by extent of failure of indentor point of any one of a number of standard testing instruments to penetrate the product.

Heat-soak test (HST) A heat-treatment which is carried out after the tempering process in order to reduce the risk of spontaneous breakage of heat treated glass in service due to nickel sulfide inclusions.

Heat strengthened glass Glass with a medium residual compressive surface stress. Heat strengthened glass is required, according to ASTM C 1048-04, to have a residual compressive surface stress between 24 MPa (3 500 psi) and 52 MPa (7 500 psi). In European standards, the fragmentation count and the maximum fragment size is specified (EN 1863-1:2000; EN 1863-2:2004).

Heat treated glass Glass that has been thermally treated to some extent. The term includes heat strengthened and fully tempered glass.

Homogeneous The opposite of heterogeneous. Consisting of the same element, ingredient, component, or phase throughout, or of uniform composition throughout.

Impact The single instantaneous stroke or contact of a moving body with another either moving or at rest.

Impact strength Measure of toughness of a material, as the energy required to break a specimen in one blow.

Inherent strength The part of the tensile strength that is *not* due to compressive residual stresses but to the resistance of the material itself. For float glass, this is approximately (even float glass has some compressive residual stresses) the measured macroscopic resistance.

Insulating glass unit (IGU) A piece of glazing consisting of two or more layers of glazing separated by a spacer along the edge and sealed to create a dead air space or a vacuum between the layers in order to provide thermal insulation. The dead air space is often filled with inert gas (argon or, less commonly, krypton).

Interlayer Very thin layer between two materials. In laminated glass: a transparent, tough plastic sheeting material, such as PVB, that is able to retain the fragments after fracture.

Intumescence The swelling and charring of materials when exposed to fire.

Joint The location at which two adherents are held together by an adhesive.

Laminated glass Two or more panes of glass bonded together with a plastic interlayer.

Lateral load Short form of 'out-of-plane load', often also used as a short form for → uniform lateral load.

Lehr Similar to an oven, used to anneal glass by reheating it and allowing it to cool slowly.

Lifetime prediction Predicting, by some engineering technique, the period of time during which a structural element will perform according to its specification.

Load duration factor The effect of a given load depends not only on its intensity, but also on the duration of a glass element's exposure to the load. This is often accounted for by applying a duration-dependent factor, the so called 'load duration factor', either to the load intensity or to some reference resistance.

Load shape Describes the geometric properties of a load, e.g. whether it is a distributed load, a point load, a line load, or a free-form load, where on a structural element it is applied and whether it is uniform, triangular or has some other shape. A complete characterization of a load must include its shape and intensity (cf. 'action intensity').

Loading time The time period during which a load is applied.

Low emissivity coating (low-e coating) A transparent metallic or metallic oxide coating that saves energy and increases comfort inside a building by reducing heat loss to the environment.

Low iron glass Extra clear glass, which has a reduced iron oxide content in order to lessen the green tinge inherent in ordinary clear float glass.

Mode I Loading condition that displaces the crack faces in a direction normal to the crack plane, also known as the opening mode of deformation.

Monotonously increasing If $x(t)$ is monotonously increasing with t, it is $x(t_2) > x(t_1)$ for any $t_2 > t_1$. The increase may or may not be linear.

Multiple-glazed units Units of three panes (triple-glazed) or four panes (quadruple-glazed) with two and three dead air spaces, respectively.

Nickel sulfide inclusion A rare, but naturally occurring impurity present in all glass that can, in certain circumstances, lead to spontaneous breakage of heat treated glass in service.

Non-uniform stress field A stress field in which the stress varies from one point of the surface to another (cf. uniform stress field).

Optical flaw depth The optically measured depth of a surface flaw.

Pane (of glass) A sheet of glass.

Predictive modelling The creation of a new model or the use of an existing model to predict the behaviour of a system, e.g. the mechanical behaviour of a structural glass element.

Profile glass Usually U-shaped, rolled glass for architectural use.

PVB (polyvinyl butyral) Polyvinyl butyral is a viscoelastic resin that is made from vinyl acetate monomer as the main raw material. It provides strong binding, optical clarity, adhesion to many surfaces, toughness and flexibility. PVB is the most commonly used interlayer material for laminated glass.

Radiation Energy released in the form of waves or particles because of a change in temperature within a gas or vacuum.

Realization Any value which the random variable X can assume is called a realization of X. 'To realize X' means to pick a value at random from its distribution.

Residual stress The residual compressive surface stress that arises from the tempering process. (The term 'prestress' is, although widely used, somewhat misleading and therefore not used in the present document.)

Rigidity The property of bodies by which they can resist an instantaneous change of shape. The reciprocal of elasticity.

Sandblasting A special glass treatment in which sand is sprayed at high velocities over the surface of the glass.

Silica Silica, also known as silicon dioxide (SiO_2), is a very common mineral composed of silicon and oxygen. Quartz and opal are two forms of silica. In nature, silica is commonly found in sand.

Silicates Silicates are minerals composed of silicon and oxygen with one or more other elements. Silicates make up about 95% of the Earth's crust.

Spacer, spacer bar Generally an aluminium bar along all edges of a double-glazed unit, filled with desiccant, which separates the two panes of glass and creates a cavity.

Strength The maximum stress required to overcome the cohesion of a material. Strength is considered in terms of compressive strength, tensile strength, and shear strength, namely the limit states of compressive stress, tensile stress and shear stress respectively.

Stress rate The stress rate $\dot{\sigma}$ is the increase in stress per unit of time, or, in other words, the derivative over time of the stress: $\dot{\sigma} = d\sigma/dt$.

Structural design The iterative process of selecting a structural element that meets a set of performance requirements that depend on the specific application. Common requirements for structural glass elements relate to aspects such as deformation, vibration, usability, aesthetics, acoustic or optical performance, and, of course, load bearing capacity.

Structural glazing Glass acting as a structural support to other parts of the building structure. It can also refer to glass that is fixed by means of bolted connectors, although the glass is not acting as a structural element in this case.

Structural sealant glazing An external glazing system in which the glass is bonded to a carrier frame without mechanical retention. Often called structural silicone glazing when a silicone adhesive/sealant is used.

Tensile strength The maximum amount of tensile stress that a material can be subjected to before failure. The definition of failure can vary according to material type, limit state and design methodology.

Thermal stress The internal stresses created when glass is subjected to variations in temperature across its area. If the temperature differentials in the glass are excessive, the glass may crack. This is referred to as thermal breakage or fracture.

Tin side The lower side of glass in the float process, i.e. the side that is in contact with the pool of molten tin.

Transient analysis An analysis that accounts for the time-dependence of input parameters.

Transparent Clear, permitting vision.

Uniaxial stress field The minor principal stress is equal to zero. An uniaxial stress field is encountered for instance in four-point-bending tests.

Uniform lateral load Uniformly distributed out-of-plane load.

Uniform stress field A stress field where the stress is equal at all points on the surface (cf. non-uniform stress field).

Viscosity A measure of the resistance of a fluid to deformation under shear stress. Viscosity describes a fluid's internal resistance to flow and may be thought of as a measure of fluid friction.

Chapter

1 Introduction

1.1 Background and motivation

The earliest man-made glass[†] objects, mainly non-transparent glass beads, are thought to date back to around 3500 BC. Such objects have been found in Egypt and Eastern Mesopotamia. Glass vases appeared in the 16th century BC, glass pots followed shortly after. The next major breakthrough was the discovery of glassblowing sometime between 27 BC and AD 14. This discovery is attributed to Syrian craftsmen. It was the Romans who began to use glass for architectural purposes, with the discovery of clear glass in Alexandria around AD 100. Cast glass windows, albeit with poor optical qualities, thus began to appear in the most important buildings. From the 11th century on, new ways of making sheet glass were developed. Panes[†] of very limited size were joined with lead strips and pieced together to create windows. Glazing[†] remained, however, a great luxury up to the late Middle Ages. The float glass[†] process, introduced commercially by Britain's Pilkington Brothers in 1959 and still used today, finally allowed large flat glass panels of very high optical quality to be produced reliably and at low cost.

For centuries, the use of glass in buildings was essentially restricted to windows and glazing. Glass building components were therefore basically required to resist out-of-plane wind loads only. Improvements in production and refining technologies such as tempering and the production of laminated glass[†] enabled glass to carry more substantial superimposed loads and therefore achieve a more 'structural' role. Architecture's unremitting pursuit of ever greater transparency has caused an increasingly strong demand for such elements ever since.

Considerable research has been undertaken in recent years to improve the understanding of the load-carrying behaviour of structural glass elements, the actions[†] those elements are exposed to and the requirements in terms of safety and serviceability that they have to meet. While these research efforts have provided much insight into the behaviour of structural glass components, the structural design of these components remains problematic. Glass failure is the consequence of the growth of surface flaws[†] under static loads. The behaviour of glass elements, therefore, strongly depends on their surface condition as well as on the environmental conditions and the loading history that they have been exposed to. The common limit state verification approach, which is mostly used for other materials and compares the maximum action effect (e. g. the maximum stress in an element during its design life[†]) with the maximum, time-independent resistance, is not readily applicable. Moreover, the extrapolation from laboratory conditions to in-service conditions is a major concern.

There is a long tradition, and therefore considerable knowledge and experience, of using rectangular glass plates to resist uniform lateral load[†]. All widely used standards, draft standards and other design methods focus on variations of this case. Unfortunately, they suffer from some notable shortcomings:

- Assumptions, limits of validity and derivations of the models are often unclear. This is mainly due to a historical mix of empirical approaches and physical models and to simplifying assumptions being deeply integrated into the models.
- Some parameters combine several physical aspects. As a result, the parameters reflect other influences than those they are supposed to and depend on the experimental setup used for their determination.
- The models contain inconsistencies. They give unrealistic results for special cases and different models yield differing results.
- Because of underlying implicit or explicit assumptions, none of the existing models takes all aspects that influence the lifetime of structural glass elements into consideration and none is applicable to general conditions. Common limitations to rectangular plates, uniform lateral loads[1], constant loads, time-independent stress distributions and the like prevent application to cases such as in-plane or concentrated loads, loads which cause not only time-variant stress levels but also time-variant stress distributions, stability problems, and connections.
- Through the condition of the specimens used to obtain strength data, an assumption about the surface condition is integrated into current design methods (homogeneous random surface flaw population; the properties vary depending on the method). The condition of the glass surface is not represented by independent design parameters that the user can modify. This is a notable drawback, especially when hazard scenarios that involve surface damage must be analysed.
- Several researchers have expressed fundamental doubts about the suitability and correctness of common glass design methods, even for the limited scope that they were developed for.

This results in an unsatisfactory situation, which has various negative consequences. Code development in Europe has reached a deadlock due to stiff opposition. The lack of confidence in 'advanced' glass models and the absence of a generally agreed design method result in frequent time-consuming and expensive laboratory testing and in inadequately designed structural glass elements. Oversized glass components are unsatisfactory from an economic and architectural point of view, undersized components are unacceptable in terms of safety.

In conclusion, the main requirements in order to be able to design and build structural glass elements with confidence are the following:

- A lifetime prediction[1] model for structural glass elements, which provides the following features, should be established:
    ◇ A comprehensive and clear derivation should allow the model, and all assumptions it is based on, to be fully understood.
    ◇ Accuracy, flexibility and scope of application should be extended with respect to current models.
    ◇ The model's parameters should each represent only one physical aspect and should be independent of test conditions.
    ◇ In hazard scenarios that involve surface damage, the main danger for a structural element arises not from the load intensity but from surface damage. A straightforward way to analyse such cases should be provided.
    ◇ A straightforward way to benefit from additional knowledge should be provided. In particular, the model should allow for less conservative design if additional data from quality control measures or research are available.
- Fundamental doubts with respect to structural glass design should be discussed and, if possible, resolved.
- Better insight should be provided into how testing should be conducted in order to obtain meaningful and safe results.

1.2 Main objectives

The current research project investigates, and endeavours to solve, the problems outlined above. In summary, this thesis aims to answer the following question: *'How can a structural glass element of arbitrary geometry be conveniently and accurately analysed and designed for general conditions?'*. More specifically, the objectives are as follows:

1. To analyse the current knowledge about, and identify research needs for, the safe and economical structural design[1] of glass elements.

2. To establish a lifetime prediction model for structural glass elements, which is as consistent and flexible as possible and offers a wide field of application. The model's parameters should each represent only one physical aspect and should be independent of test conditions. In addition to standard hazard scenarios including loads and constraint stresses, hazard scenarios that involve surface damage should be able to be analysed using the model. A comprehensive and clear derivation should allow the model, and all assumptions it is based on, to be fully understood.

3. To assess the main hypotheses of the model, discuss possibilities for its simplification and relate it to existing models. To verify the availability of the required model input data and improve existing testing procedures in order to provide more reliable and accurate model input.

4. To deploy the model to provide recommendations for the structural design of glass elements and for the laboratory testing required within the design process. To develop computer software that facilitates the application of the recommendations and enables the model to be used efficiently in research and practice.

The present study focuses on the design of glass elements for structural safety. Because the vast majority of structural glass elements are made of soda lime silica glass, only that material is considered.

The models in this work are based on quasi-static fracture mechanics. They are valid for very short loading times, but they cannot describe dynamic phenomena such as the behaviour of a glass element after fracture initiation or the response to hard impact.

1.3 Organization of the thesis

This thesis is divided into nine chapters. The organization is shown schematically in Figure 1.1.

Chapter 2 gives a short introduction to the fundamental aspects of glass as a building material. While it is by no means exhaustive, it provides the information required to understand subsequent chapters.

Chapter 3 starts with an overview of the state of knowledge with respect to laboratory testing procedures and design methods that are currently used in the context of structural glass. The main focus lies on determining the basis of the methods, on describing the simplifying assumptions they are based on, and on comparing the models to one another. The chapter closes with an analysis of present knowledge. This analysis serves to identify weaknesses and gaps in information in order to provide a focus for subsequent investigations.

Chapter 4 derives a flexible lifetime prediction model for structural glass elements based on linear elastic fracture mechanics and the theory of probability. As a prerequisite, the phenomenon of stress corrosion causing subcritical growth of surface flaws is discussed. Second, a model for the time-dependent behaviour of a single flaw is established. The last step is to extend the single-flaw model to a large number of flaws of random depth, location and orientation. This chapter focuses on fundamental concepts. Quantitative aspects are investigated in Chapter 6 and the actual use of the model is discussed in Chapter 8.

Chapter 5 assesses the main hypotheses of the model from Chapter 4 and investigates possibilities for the model's simplification. These investigations provide answers to fundamental issues such as the relevance of the risk integral and the necessity to take the crack growth threshold and the effects of biaxial stress fields into account. Furthermore, this chapter identifies the implicit assumptions behind

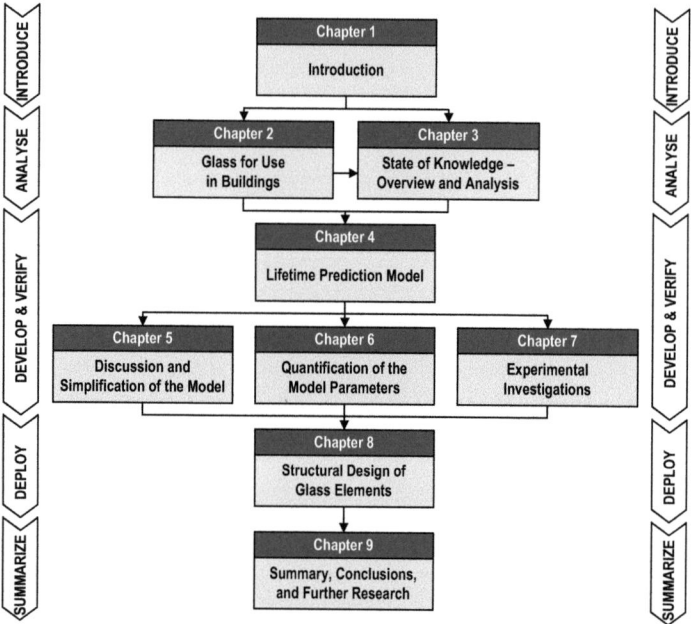

Figure 1.1: Organization of the thesis.

semi-empirical and approximate glass design methods and discusses what their range of validity is and how their parameters can (under certain conditions) be converted to the lifetime prediction model's set of fundamental parameters.

Chapter 6 verifies whether the lifetime prediction model's input parameters can be determined for in-service conditions with sufficient accuracy and reliability. To this end, a wide range of existing experimental data related to the mechanical resistance of glass are collected, compared and analysed. Furthermore, the crack growth's dependence on the stress rate[†] is assessed using the lifetime prediction model from Chapter 4 and existing large-scale testing data. Issues that need further investigation are identified. They are addressed by specifically designed experiments in Chapter 7.

Chapter 7 describes experimental investigations that were conducted in order to answer the questions that arose from Chapter 6, namely (a) Inert testing: Is inert testing feasible for structural applications? What testing procedure is suitable? Can the surface condition parameters of as-received glass[†], which were found in Section 6.5, be confirmed? (b) Deep close-to-reality surface flaws: How can deep, close-to-reality surface flaws be created for strength testing? What is their mechanical behaviour? (c) Detectability: What is the probability of detecting surface flaws through today's visual inspections?

Chapter 8 deploys the developments and findings of all preceding chapters in order to provide recommendations for the structural design of glass elements and for the laboratory testing required within the design process. To this end, the chapter commences by discussing key aspects related to the use of the lifetime prediction model for structural design. On this basis, recommendations are then developed. The chapter closes with a short presentation of the computer software that was developed to facilitate the application of the recommendations and to enable the lifetime prediction model to be used efficiently in research and practice.

Chapter 9 summarizes the principal findings and gives the most significant conclusions from the work. Finally, recommendations for future work are made.

Chapter

2

Glass for Use in Buildings

This chapter gives a short introduction to the fundamental aspects of glass as a building material. While it is by no means exhaustive, it provides the information required to understand subsequent chapters. More information can be found e. g. in Haldimann et al. (2007).

2.1 Production of flat glass

Figure 2.1 gives an overview of the most common glass production processes, processing methods and glass products. The main production steps are always similar: melting at $1600 - 1800\,°C$, forming at $800 - 1600\,°C$ and cooling at $100 - 800\,°C$.

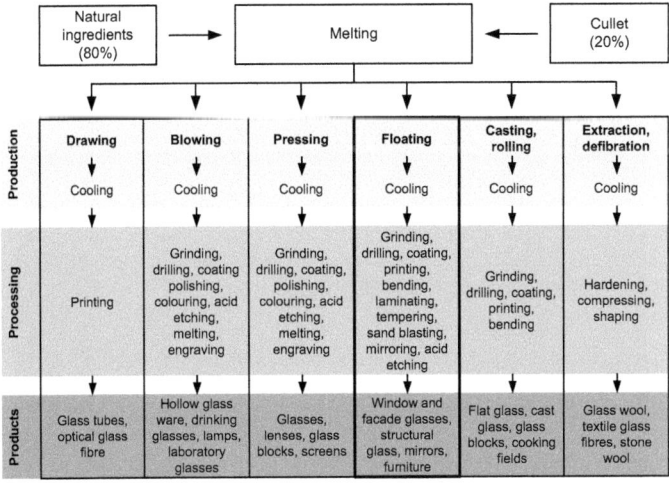

Figure 2.1: Glass production processes and products overview.

Currently the float process is the most popular primary manufacturing process and accounts for about 90% of today's flat glass production worldwide. The major advantages of this production process, introduced commercially by the Pilkington Brothers in 1959, is its low cost, its wide availability,

5

the superior optical quality of the glass and the large size of panes that can be reliably produced. The mass production process together with many post-processing and refinement technologies invented or improved over the last 50 years (see Section 2.3) have made glass cheap enough to allow it to be used extensively in the construction industry and arguably to become 'the most important material in architecture' (Le Corbusier). Within the last two decades, further progress in the field of refinement technologies (tempering, laminating) aided by structural analysis methods (e. g. finite element method) have enabled glass to be used for structural building elements.

Float glass[†] is made in large manufacturing plants that operate continuously 24 hours a day, 365 days a year. The production process is shown schematically in Figure 2.2. The raw materials are melted in a furnace at temperatures of up to 1550 °C. The molten glass is then poured continuously at approximately 1000 °C on to a shallow pool of molten tin whose oxidation is prevented by an inert atmosphere consisting of hydrogen and nitrogen. Tin is used because of the large temperature range of its liquid physical state (232 °C – 2270 °C) and its high specific weight in comparison with glass. The glass floats on the tin and spreads outwards forming a smooth flat surface at an equilibrium thickness of 6 mm to 7 mm, which is gradually cooled and drawn on to rollers, before entering a long oven, called a lehr[†], at around 600 °C. The glass thickness can be controlled within a range of 2 mm to 25 mm by adjusting the speed of the rollers. Reducing the speed increases glass thickness and vice versa. The annealing lehr slowly cools the glass to prevent residual stresses[†] being induced within the glass. After annealing[†], the float glass is inspected by automated machines to ensure that obvious visual defects[†] and imperfections are removed during cutting. The glass is cut to a typical size of 3.12 m × 6.00 m before being stored. Any unwanted or broken glass is collected and fed back into the furnace to reduce waste. At some float plants, so-called on-line coatings[†] (hard coatings) can be applied to the hot glass surface during manufacture.

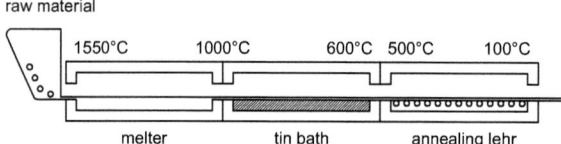

Figure 2.2:
Production process for float glass.

As a consequence of this production process, the two faces of glass sheets are not completely identical. On the tin side[†], some diffusion of tin atoms into the glass surface occurs. This may have an influence on the behaviour of the surface when it is glued (Lotz, 1995). The mechanical strength of the tin side has been found to be marginally lower than that of the air side[†]. This is not attributed to the diffused tin atoms but to the contact of the tin side with the transport rollers in the cooling area. This causes some surface defects that reduce the strength (Sedlacek et al., 1999). This interpretation is supported by the fact that the strength of intentionally damaged glass specimens has been found to be independent of the glass side (Güsgen, 1998). The tin side can be detected thanks to its bluish fluorescence when exposed to ultraviolet radiation[†].

An overview of relevant European standards for glass products is given in Appendix A.

2.2 Material properties

2.2.1 Composition and chemical properties

Glass is generally defined as an 'inorganic product of fusion which has been cooled to a rigid condition without crystallization'. The term therefore applies to all noncrystalline solids showing a glass transition. Most of the glass used in construction is soda lime silica glass (SLS). For some special applications (e. g. fire protection glazing, heat resistant glazing), borosilicate glass (BSG) is used. The latter offers a very high resistance to temperature changes as well as a very high hydrolytic and acid resistance. Table 2.3 gives the chemical composition of these two glass types according to current European standards. In contrast to most other materials, glasses do not consist of a geometrically

2.2. MATERIAL PROPERTIES

regular network of crystals, but of an irregular network of silicon and oxygen atoms with alkaline parts in between (Figure 2.4).

Table 2.3:
Chemical composition of soda lime silica glass and borosilicate glass; indicatory values (mass %) according to (EN 572-1:2004) and (EN 1748-1-1:2004).

			Soda lime silica glass	Borosilicate glass
Silica sand	SiO_2		69 – 74%	70 – 87%
Lime (calcium oxide)	CaO		5 – 14%	–
Soda	Na_2O		10 – 16%	0 – 8%
Boron-oxide	B_2O_3		–	7 – 15%
Potassium oxide	K_2O		–	0 – 8%
Magnesia	MgO		0 – 6%	–
Alumina	Al_2O_3		0 – 3%	0 – 8%
others			0 – 5%	0 – 8%

Figure 2.4:
Schematic view of the irregular network of a soda lime silica glass.

○ oxygen (O)
● silicone (Si)
(Na) sodium (Na)
(Ca) calcium (Ca)

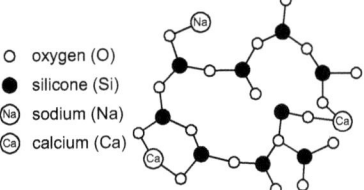

The chemical composition has an important influence on the viscosity', the melting temperature T_S and the thermal expansion coefficient α_T of glass. While the melting temperature is about 1710 °C for pure silica' oxide, it drops to 1300 °C – 1600 °C through the addition of alkali'. The thermal expansion coefficient is about $0.5 \cdot 10^{-6} \, K^{-1}$ for pure silica glass and $9 \cdot 10^{-6} \, K^{-1}$ for soda lime silica glass. During the cooling of the liquid glass, its viscosity increases constantly until solidification at about 10^{14} Pa s. The temperature at solidification is called transformation temperature T_g and is about 530 °C for soda lime silica glass. In contrast to crystalline materials, the transition between liquid and solid state does not take place at one precise temperature but over a certain temperature range. The properties thus change gradually. The glass is actually 'freezing' and no crystallization takes place. The 'super-cooled liquid' nature of glass means that, unlike most solids, the electrons in glass molecules are strictly confined to particular energy levels. Since this means that the molecules cannot alternate between different states of excitement by absorbing radiation in the bandwidths of visible and near-infrared, they do not absorb or dissipate those forms of radiant energy. Instead, the energy passes straight through the molecules as if they were not there. However, due to unavoidable impurities in the soda-lime-silica mix, typical window glass does absorb some radiation that might otherwise pass through (cf. Section 2.2.2). Small amounts of iron oxides are responsible for the characteristic greenish colour of soda lime silica glass (e.g. Fe^{2+}: blue-green; Fe^{3+}: yellow-brown).

Tinted glass is produced by adding metal oxides (iron oxide, cobalt oxide, titanium oxide and others). Standard colours are bronze, grey, green and pink. As the colour is very sensitive even to little changes of the glass composition, an exact colour match between different production lots is difficult to obtain.

One of the most important properties of glass is its excellent chemical resistance to many aggressive substances, which explains its popularity in the chemical industry and makes glass one of the most durable materials in construction.

Table 2.5: Physical properties of soda lime silica and borosilicate glass (EN 1748-1-1:2004; EN 572-1:2004).

			Soda lime silica glass	Borosilicate glass
Density	ρ	kg/m^3	2 500	2 200 – 2 500
Knoop hardness[†]	HK$_{0,1/20}$	GPa	6	4.5 – 6
Young's modulus	E	MPa	70 000	60 000 – 70 000
Poisson's ratio	ν	–	0.23[*]	0.2
Coefficient of thermal expansion[†]	α_T	$10^{-6}\,\text{K}^{-1}$	9	Class 1: 3.1 – 4.0
				Class 2: 4.1 – 5.0
				Class 3: 5.1 – 6.0
Specific thermal capacity	c_p	J kg^{-1} K^{-1}	720	800
Thermal conductivity	λ	W m^{-1} K^{-1}	1	1
Average refractive index within the visible spectrum[‡]	n	–	1.52[§]	1.5
Emissivity[†] (corrected[¶])	ε	–	0.837	0.837

[*] The code gives 0.2. In research and application, values between 0.22 and 0.24 are commonly used.
[†] Mean between 20 °C and 300 °C.
[‡] The refractive index is a constant for a given glazing material, but depends on the wavelength. The variation being small within the visible spectrum, a single value provides sufficient accuracy.
[§] The code gives a rounded value of 1.50.
[¶] For detailed information on the determination of this value see EN 673:1997.

2.2.2 Physical properties

The most important physical properties of soda lime silica and borosilicate glass are summarized in Table 2.5. Optical properties depend on the glass thickness, the chemical composition and the applied coatings. The most evident property is the very high transparency within the visible range of wavelengths ($\lambda \approx 380\,\text{nm} - 750\,\text{nm}$). Whilst the exact profiles of the non-transmitted (i. e. absorbed and reflected) radiation spectrum varies between different types of glass, they are usually in the wavelengths outside the visible and near infrared band. Due to interaction with O_2-ions in the glass, a large percentage of UV radiation is absorbed. Long-wave infrared radiation ($\lambda > 5\,000\,\text{nm}$) is blocked because it is absorbed by Si-O-groups. This is at the origin of the greenhouse effect: visual light passes through the glass and heats up the interior, while emitted long-wave thermal radiation is unable to escape. With its refractive index of about 1.5, the reflection of visual light by common soda lime silica glass is 4% per surface which gives a total of 8% for a glass pane. This reduces transparency but can be avoided by applying special coatings.

At room temperature, the dynamic viscosity of glass is about $10^{20}\,\text{Pa s}$. (For comparison, the viscosity of water is $10^{-1}\,\text{Pa s}$ and of honey, $10^5\,\text{Pa s}$.) Given this extremely high viscosity at room temperature, it would take more than the earth's age for 'flow' effects to become visible to the naked eye. Although the notion of flowing glass has been repeatedly propagated, 'flow' of the glass is therefore very unlikely to be the cause of window glasses in old churches being thicker at the bottom than at the top. More realistic reasons are the poor production quality of these old glasses and surface corrosion effects caused by condensed water accumulating at the bottom of glass panes and leading to an increase in volume.

Glass shows an almost perfectly elastic, isotropic behaviour and exhibits brittle fracture. It does not yield plastically, which is why local stress concentrations are not reduced through stress redistribution as it is the case for other construction materials such as steel. The theoretical tensile strength[†] (based on molecular forces) of glass is exceptionally high and may reach 6 000 MPa – 10 000 MPa. It is however of no practical relevance for structural applications. The actual tensile strength, the relevant property for engineering, is much lower. The reason is that the surface of glass panes contains a large number of mechanical flaws of varying severity which are not necessarily visible to the naked eye. As with all brittle materials, the tensile strength of glass depends very much on these surface flaws. A glass element fails as soon as the stress intensity due to tensile stress at the tip of one flaw reaches its critical value. Flaws grow with time when loaded, the crack velocity being a function of several parameters and extremely variable. This is discussed in detail in Chapter 4. For the moment, it shall only be pointed out that the tensile strength of glass is not a material constant, but it depends on many aspects, in particular on the condition of the surface, the size of the glass element, the action history† (intensity and duration), the residual stress and the environmental conditions. The higher the stress, the longer the load duration and the deeper the initial surface flaw, the lower the effective tensile strength.

As surface flaws do not grow or fail when in compression, the compressive strength of glass is much larger than the tensile strength. It is, however, irrelevant for virtually all structural applications. In the case of stability problems, tensile stresses develop due to buckling. At load introduction points, the Poisson's ratio effect causes tensile stresses. An element's tensile strength is, therefore, exceeded long before the critical compressive stresses are reached.

2.3 Processing and glass products

2.3.1 Introduction

Once manufactured, flat glass is often processed further to produce glass products of the shape, performance and appearance that are required to meet particular needs. This secondary processing may include:

- cutting to remove edge damage and to produce the desired pane shape and size
- edge working (arrissing, grinding, polishing) and drilling
- curving
- application of coatings
- thermal treatment to get heat strengthened or fully tempered glass† (tempering)
- heat soaking to reduce the potential for nickel sulfide-induced breakages in use
- laminating for enhanced post-breakage performance, safety on impact, bullet resistance, fire resistance or acoustic insulation
- surface modification processes for decoration, shading or privacy
- insulating glass unit† assembly to reduce heat loss and, if suitably configured, to reduce solar gain and enhance acoustic performance.

Many of these methods of glass processing are of no immediate relevance for the present work and will not be further discussed. Some, however, are relevant and shall therefore be outlined in the following sections.

The term *glass pane* will hereafter be used to refer to a single pane of sheet glass. A glass pane may be used as a monolithic glass or it may be part of an insulating glass unit, a laminated glass or some other glass assembly (Figure 2.6). *Glass unit* is a generic term for any of these.

Figure 2.6:
Basic types of glass units.

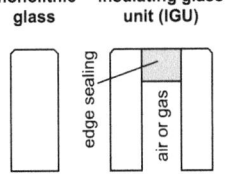
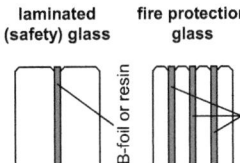

2.3.2 Tempering of glass

Principle and main effects

For structural glass applications, tempering (heat treatment) is the most important processing method. The idea is to create a favourable residual stress field featuring tensile stresses in the core of the glass and compressive stresses on and near the surfaces. The glass core does not contain flaws and therefore offers good resistance to tensile stress. The unavoidable flaws on the glass surface can only grow if they are exposed to an effective tensile stress. As long as the tensile surface stress due to actions is smaller than the residual compressive stress, there is no such effective tensile stress and consequently no crack growth (Figure 2.7).

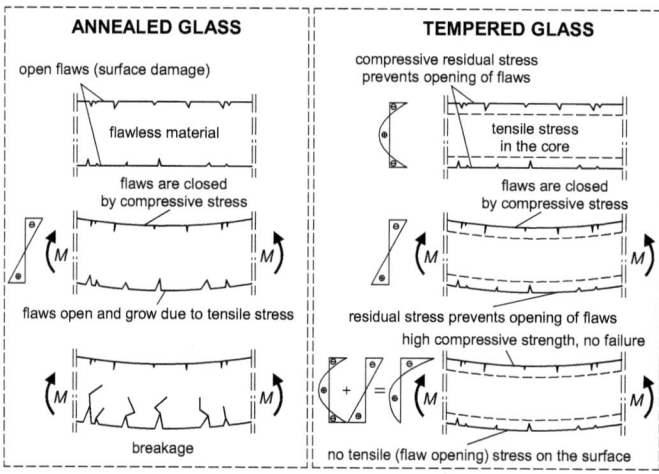

Figure 2.7: The principle of glass tempering (adapted from Sedlacek et al. (1999)).

The fracture pattern is a function of the energy stored in the glass, i.e. of the residual stress and the stress due to loads. As an example, Figure 2.8 shows the fracture pattern of specimens loaded in a coaxial double ring test setup. *Fully tempered glass* has the highest residual stress level and usually breaks into small, relatively harmless dice of about 1 cm². This fracture pattern is why fully tempered glass is also called *'safety glass'*. The term may, however, be misleading. When falling from a height of several meters, even small glass dice can cause serious injury. While fully tempered glass has the highest structural capacity of all glass types, its post-failure performance is poor due to the tiny fragments. *Heat strengthened glass*[1] provides an interesting compromise between fairly good structural performance and a sufficiently large fragmentation pattern for good post-failure performance. *Annealed glass* is standard float glass without any tempering. It normally breaks into large fragments. If, however, it is exposed to high (especially in-plane) loads, the elastic energy stored in the material due to elastic deformation[1] can lead to a fracture pattern similar to heat treated glass[1].

On an international level, no specific terminology for the different glass types has to date gained universal acceptance. In the present document, the terms from ASTM E 1300-04 are used (Table 2.9). They are widely used and tend, in the opinion of the author, to be less susceptible to misunderstandings than others.

2.3. PROCESSING AND GLASS PRODUCTS

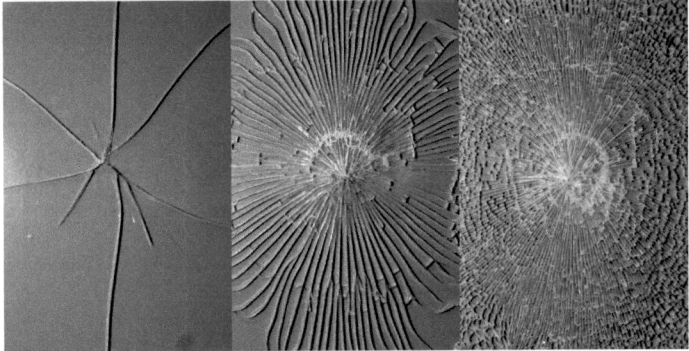

Figure 2.8: Comparison of the fracture pattern: annealed glass (left), heat strengthened glass (middle), fully tempered glass (right).

Table 2.9: Glass type terminology overview.

Level of residual surface compression	Terminology in the present document	Other frequently used terms
(almost) none	annealed glass (ANG)	float glass
medium	heat strengthened glass (HSG)	partly toughened glass;
high	fully tempered glass (FTG)	tempered glass; (thermally) toughened glass
unspecified (HSG or FTG)	heat treated glass	

Fully tempered glass

During the thermal tempering process, float glass is heated to approximately 620 − 675 °C (approximately 100 °C above the transformation temperature) in a furnace and then quenched (cooled rapidly) by jets of cold air. This has the effect of cooling and solidifying first the surface and then the interior of the glass. Within the first seconds, the cooling process results in tensile stresses on the surface and compressive stresses in the interior. As the glass is viscous in this temperature range, the tensile stresses can relax rapidly. If the starting temperature is too low, the relaxation cannot take place and the tensile stresses may cause the glass to shatter in the furnace. As soon as the temperature on the glass surface falls below T_g (approx. 525 °C), the glass solidifies and relaxation stops immediately. The temperature distribution is approximately parabolic, the interior being hotter at this stage. Finally, the interior cools as well. As its thermal shrinkage is resisted by the already solid surface, the cooling leads to the characteristic residual stress field with the surfaces being in compression and the interior in tension. To obtain an optimal result with maximum temper stress, the process has to be managed so that the surface solidifies exactly at the moment when the maximum temperature difference occurs and the initial tensile stress has relaxed. Borosilicate glass is difficult to temper by high air pressure or even by quenching in liquids because of its low thermal expansion coefficient.

The typical residual compressive surface stress varies between 80 MPa and 150 MPa for fully tempered soda lime silica glass. According to ASTM C 1048-04, it is required to have either a minimum surface compression of 69 MPa (10 000 psi) or an edge compression of not less than 67 MPa (9 700 psi). In European standards, the fragmentation count and the maximum fragment size is specified (EN 12150-1:2000; EN 12150-2:2004).

Fairly accurate numerical modelling of the tempering process is possible (Bernard et al., 2002; Laufs, 2000b; Schneider, 2001). This is especially helpful to estimate tempering stresses for more complex geometries such as boreholes. The most important parameters of the tempering process are the glass thickness, the thermal expansion coefficient of the glass and the heat transfer coefficient between glass and air. In particular the heat transfer coefficient is often difficult to estimate. It depends on the quenching (jet geometry, roller influence, air pressure, air temperature, etc.) and therefore varies widely for different glass manufacturers.

Heat strengthened glass

Heat strengthened glass is produced using the same process as for fully tempered glass, but with a lower cooling rate. The residual stress and therefore the tensile strength is lower. The fracture pattern of heat strengthened glass is similar to annealed glass, with much bigger fragments than for fully tempered glass. Used in laminated glass elements, this large fracture pattern results in a significant post-breakage structural capacity.

As the stress gradient depends on the glass thickness and the glass must be cooled down slowly, thick glasses (> 12 mm) cannot be heat strengthened using the normal tempering process.

ASTM C 1048-04 requires that heat strengthened glass has a residual compressive surface stress between 24 MPa (3 500 psi) and 52 MPa (7 500 psi). In European standards, the fragmentation count and the maximum fragment size is specified (EN 1863-1:2000; EN 1863-2:2004).

Chemical tempering

Chemical tempering is an alternative tempering process that does not involve thermic effects and produces a different residual stress profile. Cutting or drilling remains possible, even after tempering. In structural applications, chemical tempering is extremely rare. It is used for special geometries where usual tempering processes cannot be applied, e. g. glasses with narrow bending angles. The process is based on the exchange of sodium ions in the glass surface by potassium ions, which are about 30% bigger. Only a very thin zone at the glass surface is affected (Figure 2.10). The actual depth of the compression zone is time-dependent (about 20 µm in 24 h) (Wörner et al., 2001). If surface flaws are deeper than the compression zone, their tip is in the zone of tensile stress and subcritical crack growth occurs without external load. This phenomenon, known as self-fatigue, can cause spontaneous failure, even of glass elements that have never been exposed to external loads. For a fracture mechanics investigation, see Bakioglu et al. (1976). An improved chemical tempering process is currently being developed, see e. g. Abrams et al. (2003); Sglavo et al. (2004); Sglavo (2003). While the scatter of the strength can be reduced, the problem of self fatigue persists and the process is expensive.

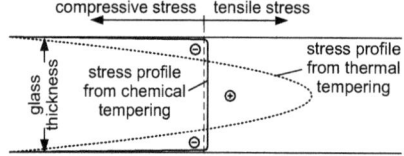

Figure 2.10:
Comparison of the stress profiles obtained by thermal and chemical tempering.

Tolerances and practical aspects

An attempt to work heat treated glass usually causes it to shatter immediately. Any cutting, drilling or grinding must therefore be carried out *before* the glass is tempered.

The heating of the glass to more than the transformation temperature and the fixing in the furnace causes some deformation. It depends on the furnace and the glass thickness, but generally increases with increasing aspect ratio[1] of a glass element. This can limit the feasible slenderness of glass beams. Furthermore, geometric tolerances are considerably higher than those of annealed glass. In particular, edges and holes in laminated glass elements made of heat treated glass are generally not flush. This cannot be corrected by grinding (see above) and must therefore be accounted for by well thought-out

2.3. PROCESSING AND GLASS PRODUCTS

details and connections. Finally, the deformation often reduces the optical quality of heat treated glass.

Specialized glass processing firms are able to temper bent glasses, but various limitations on radii and dimensions may apply.

Nickel sulfide-induced spontaneous failure

Fully tempered glass elements have a small but not negligible risk of breaking spontaneously within a few years of production. At the origin of such spontaneous failures are nickel sulfide (NiS) inclusions[1] (Figure 2.11) that cannot be avoided completely during production. Under the influence of temperature, such NiS particles can increase in volume by about 4% due to a phase change. This expansion in combination with the high tensile stresses in the glass core due to thermal tempering can cause spontaneous failure.

Figure 2.11:
Microscopic image of a nickel-sulfide inclusion in fully tempered glass (courtesy of MPA Darmstadt, Germany).

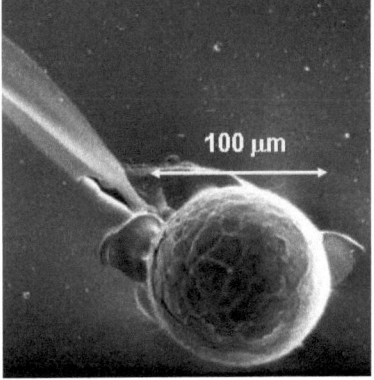

The risk of spontaneous failure due to inclusions can be significantly reduced, but not totally eliminated[1], by the heat-soak test[1]. This test consists in slowly heating up the glass and maintaining a certain temperature for several hours. This accelerates the phase change, and glass elements containing dangerous inclusions fail during the test. Depending on the location, client and glass processor involved, the heat-soak test is performed according to DIN 18516-4:1990, EN 14179-1:2005 or the German building regulation BRL-A 2005. All three regulations specify a holding temperature of $290 \pm 10\,°C$. The duration of the holding period is 8 h according to DIN 18516-4:1990, 4 h according to BRL-A 2005 and 2 h according to EN 14179-1:2005.

2.3.3 Laminated glass

Laminated glass consists of two or more panes of glass bonded together by some transparent[1] plastic interlayer[1]. The glass panes may be equal or unequal in thickness and may be the same or different in heat treatment. The most common lamination process is autoclaving at approx. $140\,°C$. The heat and the pressure of up to 14 bar ensure that there are no air inclusions between the glass and the interlayer.

Laminated glass is of major interest in structural applications. Even though tempering reduces the time dependence of the strength and improves the structural capacity of glass, it is still a brittle material. Lamination of a transparent plastic film between two or more flat glass panes enables a significant improvement of the post breakage behaviour: after breakage, the glass fragments adhere to the film so that a certain remaining structural capacity is obtained as the glass fragments

---

[1] According to EN 14179-1:2005, there is at most one failure in 400 t of heat soaked glass.

'arch' or lock in place. This capacity depends on the fragmentation of the glass and increases with increasing fragment size (Figure 2.12). Therefore, laminated glass elements achieve a particularly high remaining structural capacity when made from annealed or heat strengthened glass that breaks into large fragments. The post-breakage behaviour furthermore depends on the interlayer material.

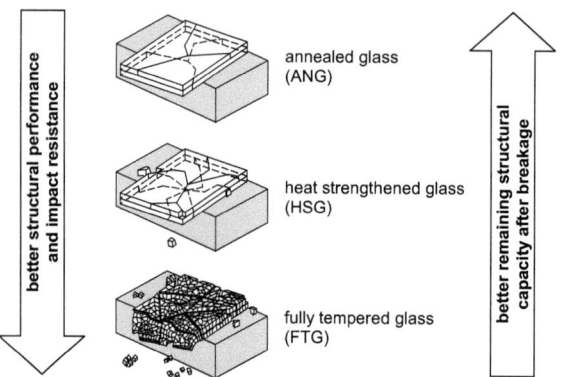

Figure 2.12:
Post breakage behaviour of laminated glass made of different glass types (adapted from Sedlacek et al. (1999)).

Today's most common interlayer material is polyvinyl butyral[†] (PVB). Because PVB blocks UV radiation almost completely, PVB foils are sometimes also called UV-protection-foils. The nominal thickness of a single PVB foil is 0.38 mm. Normally, two (0.76 mm) or four (1.52 mm) foils form one PVB interlayer. For thick or heat treated glasses, up to six may be appropriate. PVB is a viscoelastic material, i. e. its physical properties depend strongly on the temperature and the load duration. At room temperature, PVB is comparatively soft with an elongation[†] at breakage of more than 200%. At temperatures well below 0°C and for short loading times[†], PVB is in general able to transfer shear stress. For higher temperatures and long loading times, the shear transfer is greatly reduced.

Alternative transparent interlayer materials have recently been developed with the aim of achieving higher stiffness, temperature resistance, tensile strength or resistance to tearing. A well known example is DuPont's SentryGlas® Plus (Bennison et al., 2002; DuPont 2003; Pilkington Planar 2005). The high stiffness can make the lamination of such interlayers difficult.

In addition to the transparent interlayers, coloured or printed ones are also available. Other materials, i. e. transparent resins with 1 mm to 4 mm layer thickness, are sometimes used to achieve special properties such as sound insulation or to include functional components such as solar cells or LEDs.

Fire protection glass is laminated glass with one or more special transparent intumescent[†] interlayer(s). When exposed to fire, the pane facing the flames fractures but remains in place and the interlayers foam up to form an opaque insulating shield that blocks the heat of the blaze.

Bullet-resistant and blast-resistant glasses are laminated glasses using various impact energy absorbing interlayers.

Chapter

3

State of Knowledge – Overview and Analysis

This chapter starts with an overview of the state of knowledge with respect to laboratory testing procedures and design methods that are currently used in the context of structural glass. The main focus lies on determining the basis of the methods, on describing the simplifying assumptions they are based on, and on comparing the models to one another. The chapter closes with an analysis of present knowledge. This analysis serves to identify weaknesses and gaps in information in order to provide a focus for subsequent investigations.

3.1 Laboratory testing procedures

Numerical modelling is closely related to testing. Design methods cannot, therefore, be discussed independently of the corresponding testing procedures. It is therefore pertinent to discuss briefly the more common testing procedures.

This section focuses on the testing procedures, while the reliability of the results is assessed in Chapter 6.

3.1.1 Static fatigue tests

Static long-term tests with constant stress, often called '*static fatigue tests*'[1], are usually performed using a four point bending test setup (4PB). A *constant load* is applied to the specimens and the time to failure is measured. The advantage of such tests is that the test conditions are similar to the in-service conditions of structural glass elements carrying mainly dead loads. A major disadvantage is that such tests are extremely time-consuming. If a specimen's surface condition or the stress corrosion behaviour differs only slightly from the assumptions used to design the test, the specimen may only fail after several years.

3.1.2 Dynamic fatigue tests

The term '*dynamic fatigue test*' is a generic term used for constant load rate testing, for constant stress rate testing and for testing with cyclic loading. It is mostly performed using a *four point bending* or *coaxial double ring* (CDR) test setup. The latter is also known as *concentric ring-on-ring* test setup. Figure 3.1 shows a schematic representation of the two test setups.

---

[1] The term 'fatigue test' is somewhat misleading because engineers usually associate 'fatigue' with cyclic loading. In the context of glass it is used because the subcritical crack growth caused by stress corrosion is often called 'static fatigue'.

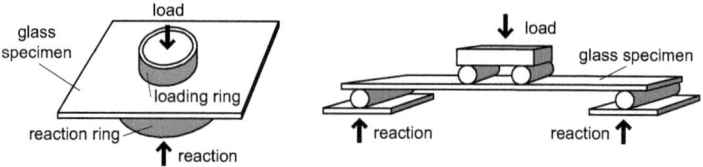

Figure 3.1: Schematic representation of coaxial double ring (left) and four point bending (right) test setups.

In 4PB tests, the specimen is exposed to an approximately uniaxial stress field[†] ($\sigma_1 \neq 0$, $\sigma_2 = 0$). In CDR tests, an equibiaxial stress field[†] ($\sigma_1 = \sigma_2$) is obtained[2]. The failure stress is a function of the stress rate. Both test setups are simple and provide short times to failure even for specimens with small surface defects (e. g. as-received glass). On logarithmic scales, the failure stress as a function of the stress rate has a slope of $1/(n+1)$. If v_0 is constant, this allows for the determination of the exponential crack velocity parameter n from tests with different stress rates. There are two main uses for dynamic fatigue testing:

- Measurement of *failure stresses* at ambient conditions (details see below). Testing is mostly done using as-received glass specimens or specimens with artificially induced homogeneous surface damage. The data obtained represent a combination of the specimen's surface condition and the crack growth behaviour during the tests.
- Measurement of *crack velocity parameters* by direct observation of surface cracks. Since accurate results can only be obtained if the flaw characteristics are well known, indentation cracks are often used.[3]

In Europe, the testing procedure that is mostly used to obtain glass material data is the *coaxial double ring test*. It is standardized in EN 1288-1:2000 (fundamentals), EN 1288-2:2000[4] (R400 test setup) and EN 1288-5:2000 (R45 and R30 test setups). Details on the different setups are given in Table 3.2. Another common procedure, the *four point bending test*, is standardized in EN 1288-3:2000. In all these tests, the stress rate to be used is 2 ± 0.4 MPa/s.

Table 3.2: Coaxial double ring test geometries in European standards.

Test setup	Standard	Loading ring radius (mm)	Reaction ring radius (mm)	Tested area* (mm²)	Specimen edge length (mm)
EN CDR R45	EN 1288-5:2000	9	45	254	100 × 100 (±2)
EN CDR R400	EN 1288-2:2000	300 ± 1	400 ± 1	240 000[†]	1 000 × 1 000 (±4)

* This is the surface area in uniform, equibiaxial tension = the area inside the loading ring (exception, see [†]).
[†] This is the value from the code. It does *not* correspond to the area within the load ring (282 743 mm²).

Statistical analysis of the test results is generally done by fitting a two-parameter Weibull distribution (Weibull, 1939, 1951) to the experimental failure stress data:

$$P_f(\sigma_{f,A}) = 1 - \exp\left[-\left(\frac{\sigma_{f,A}}{\theta_A}\right)^\beta\right] \quad (3.1)$$

$P_f(\sigma_{f,A})$ cumulative probability of failure

$\sigma_{f,A}$ breakage stress of specimens of which the surface area A is exposed to tensile stress

---

[2] For detailed information on the CDR testing procedure, the interested reader should refer to seminal work on the subject such as Schmitt (1987) (basis for EN 1288-2:2000, in German) or Simiu et al. (1984).
[3] For details on the procedure, see e. g. Green et al. (1999); Sglavo et al. (1997); Sglavo and Green (1995, 1999).
[4] DIN 52292-2:1986, which was used for the majority of tests performed in the past, has been replaced by this standard. Apart from the suppression of the test setup R200, which was hardly ever used, it does not contain any relevant changes.

3.2. DESIGN METHODS

θ_A scale parameter of the Weibull distribution (depends on A)

β shape parameter of the Weibull distribution

Various methods for parameter estimation exist, see Section E.3. The procedure standardized in EN 12603:2002 was often used in the past.[5] It is based on point estimates and the median-rank based empirical failure probability given in Equation (E.6). For details on this approach as well as on alternative methods, see Section E.3.

3.1.3 Direct measurement of the growth of large through-thickness cracks

Particularly before measurements on indentation cracks (cf. Section 3.1.2) became popular, this experimental approach was used to determine crack velocity parameters. The growth of a large through-thickness crack is directly measured (e. g. optically or using sound waves) as a function of the stress intensity factor. On one hand, this is a direct and relatively precise approach. On the other hand however, engineers designing structural glass elements are not interested in the behaviour of such large through-thickness cracks.

3.1.4 Tests for the glass failure prediction model

The glass failure prediction model (GFPM) does not use the above mentioned standard procedures. The two interdependent surface flaw parameters $\tilde{m}$ and $\tilde{k}$ are determined by loading rectangular glass plates with a uniform lateral load. The visually determined failure origin, the stress history at the failure origin and a rather complex iterative procedure are used to find the parameters (see Section 3.2.7). Only one crack velocity parameter is explicitly considered in the GFPM: The value of 16 in Equation (3.22) is actually the exponential crack velocity parameter n.

3.2 Design methods

3.2.1 Introduction

The current situation concerning design methods for structural glass elements has been outlined in Chapter 1. Clearly, before investigations aiming at generalizing and improving current methods can be undertaken, a comprehensive overview of the current state of knowledge must be established. This is what this section focuses on. It aims not so much at explaining every single detail related to the application of the methods, but at working out their basis. This section also describes the simplifying assumptions the methods are based on and how they compare with one another. In addition to providing an in-depth understanding, this section aims to identify weaknesses and gaps in information that require further investigations (Section 3.3).

For ensuing parts of the present chapter, basic knowledge of the phenomena of stress corrosion and subcritical crack growth is very useful. Readers who do not feel familiar with these topics might want to read Section 4.1 first.

3.2.2 Allowable stress based methods

Despite the inaccuracy of this over-simplistic approach and the fact that the concept of allowable stress is rarely used in current building design standards, allowable stress design methods are still widely used to design glass elements. It is mainly the extreme ease of use and the simplicity of these methods that make them attractive. The general verification format is:

$$\sigma_E \leq \sigma_{\text{adm}} \tag{3.2}$$

---

[5] The older German national standard DIN 55303-7:1996, which was used for many research projects and publications, is essentially equivalent.

σ_E maximum in-plane principal stress, calculated using the characteristic values of the actions of the most unfavourable design situation

σ_{adm} allowable principal in-plane stress (the fracture strength found in experiments, divided by a global safety factor that accounts for all uncertainties and variances associated with actions, resistance and modelling)

There is no way of considering the effects of the element's size, the environmental conditions, the duration of load and the like, or of taking a specific target failure probability into account. These aspects must all be somehow 'included' in the recommended σ_{adm} values.

The German technical guidelines TRLV 1998 and TRAV 2003 are well known and widely used examples of design guides based on allowable stresses. Both documents apply to glass panes exposed to uniform lateral loads only. The recommended allowable stresses for static loads are summarized in Table 3.3. For impact loads, TRAV 2003 gives the following allowable stresses: 80 MPa for ANG, 120 MPa for HSG, 170 MPa for FTG. Additionally, both guidelines contain a series of more detailed specifications on how to account for load combinations, they give special requirements that must be met and modified allowable stress values for a series of specific situations.

Allowable stresses have also been proposed for edges of glass beams, e. g. by Hess (2000), Güsgen (1998) and Laufs (2000b).

	Allowable stress σ_{adm} (MPa)	
	vertical glazing	overhead glazing
annealed glass (ANG)	18	12
fully tempered glass (FTG)	50	50
laminated ANG	22.5	15 (25*)

Table 3.3: Allowable stresses for glass panes exposed to uniform lateral load according to TRLV 1998 and TRAV 2003.

* only for the lower glass pane in the hazard scenario 'upper pane broken'

Allowable stress based design methods have some notable drawbacks:
- They do not account for the actual physical phenomena that govern the mechanical behaviour of glass.
- Scatter and uncertainty of the influencing parameters differ. With only one global safety factor, this cannot be accounted for.
- The approach is only valid for situations that conform to small displacement theory. It is e. g. not suitable for cases of geometric non-linearity.

3.2.3 DELR design method

Verification format

The design method of damage equivalent load and resistance, called DELR design method hereafter, was the first European glass design method that tried to account for the specific behaviour of glass in an adequate and transparent way. It is compatible with the current generation of standards based on partial safety factors. Presented to a larger public in Sedlacek et al. (1999), the design method is based on research work by Richter (1974), Kerkhof (1977); Kerkhof et al. (1981), Exner (1983, 1986), Blank (1993), Güsgen (1998) and others. Originally developed for glass plates, it was also extended to cover glass beams. The maximum principal design stress $\sigma_{max,d}$ is compared to an equivalent resistance as follows:

$$\sigma_{max,d} \leq \frac{\sigma_{bB,A_{test},k}}{\alpha_{\sigma}(q,\sigma_V) \cdot \alpha(A_{red}) \cdot \alpha(t) \cdot \alpha(S_v) \cdot \gamma_{M,E}} + \frac{\sigma_{V,k}}{\gamma_{M,V}} \qquad (3.3)$$

$\alpha_{\sigma}(q,\sigma_V)$ coefficient to account for the stress distribution on the glass surface; $q =$ uniform lateral load[6], $\sigma_V =$ residual surface stress due to tempering

---
[6] Sedlacek et al. (1999) uses p. q is used here for compatibility with the rest of the document.

3.2. DESIGN METHODS

$\alpha(A_{red})$ coefficient to account for the size of the decompressed surface[†] area A_{red} (for annealed glass, A_{red} is equal to the entire surface area)

$\alpha(t)$ coefficient to account for the load duration

$\alpha(S_v)$ coefficient to account for load combination and environmental conditions

$\sigma_{max,d}$ design value of the maximum in-plane principal stress in the element, calculated according to current action standards[7]

$\sigma_{bB,A_{test},k}$ characteristic value of the inherent bending fracture strength in R400 coaxial double ring tests according to EN 1288-2:2000 (see Section 3.1.2; 5% fractile, confidence level 0.95, surface area[8] $A_{test} = 0.24\,m^2$, stress rate $= 2 \pm 0.4\,MPa/s$)

$\sigma_{V,k}$ characteristic value (5% fractile) of the absolute value (compression = positive) of the residual surface stress (normally induced by thermal or chemical tempering; called 'prestress' in the DELR design method;)

$\gamma_{M,E}$ partial factor for the inherent strength[†]

$\gamma_{M,V}$ partial factor for the residual stress

Coefficients

A set of coefficients is used to compensate for the differences between laboratory test conditions (used to determine the strength) and actual in-service conditions. The *non-homogeneous stress distribution* on the glass surface is accounted for as follows:

$$\alpha_\sigma(q, \sigma_V) = \left[\frac{1}{A_{red}} \int_{A_{red}} \left(\frac{\sigma_1(x,y)}{\sigma_{max,d}} \right)^\beta dx dy \right]^{1/\beta} \quad (3.4)$$

$\sigma_1(x,y)$ is the major principal stress at the point (x,y) on the surface and depends (like A_{red}) on σ_V. The Weibull shape parameter β is assumed to be 25. This value has been defined by Blank (1993) based on experiments on float glass samples with artificially induced homogeneous surface damage (sandblasting'). It does not directly reflect the test data, but reflects a so-called 'limiting distribution' that lies somewhat below the test data. For standard cases, tabulated values of α_σ (from finite element calculations) are given, a simple but conservative assumption is $\alpha_\sigma = 1.0$.

The *size effect* is accounted for as follows:

$$\alpha(A_{red}) = \left(A_{red}/A_0 \right)^{1/\beta} \quad (3.5)$$

The coefficient $\alpha(t)$ accounts for the *load duration*. It depends on the subcritical crack growth, the duration of all loads in a load combination, the overlapping probability of wind- and snow load, the bending strength determined in tests, the stress rate used in these tests, the surface area and the required lifetime. For usual conditions and a design life of 50 years, Sedlacek et al. (1999) proposes to use $\alpha(t) = 3.9$.

The coefficient $\alpha(S_v)$ takes the *relative magnitude of the different loads* within a load combination as well as *environmental conditions* into account. Its calculation is complex and too lengthy to be discussed here; the interested reader should refer to Sedlacek et al. (1999). The difference with respect to other design methods is that two sets of crack velocity parameters are used in the calculation of $\alpha(S_v)$: one for 'winter conditions' ($S_{Winter} = 0.82\,m/s(MPa\,m^{0.5})^{-n}$) and one for 'summer conditions' ($S_{Sommer} = 0.45\,m/s(MPa\,m^{0.5})^{-n}$). $n = 16$ is assumed for both conditions.

---

[7] In particular EN 1990:2002; EN 1991-1-1:2002; EN 1991-2-3:1996; EN 1991-2-4:1995; EN 1991-2-5:1997; EN 1991-2-7:1998 in connection with the National Application Documents in Europe; SIA 260:2003 (e) and SIA 261:2003 (e) in Switzerland.

[8] Sedlacek et al. (1999) uses A_0. This symbol is avoided here because is has a different meaning in the present document (unit surface area).

Partial factors

In Sedlacek et al. (1999), a partial resistance factor of $\gamma_M \approx 1.80$ is proposed for structures of medium importance. This factor is chosen by a rather particular approach involving two Weibull distributions: First, a 'characteristic value of the inherent bending strength' $\sigma_{bB,A_{test},k} = 45\,\text{MPa}$ is defined as the 5% fractile of a Weibull distribution with the parameters $\theta_{A_{test}} = 74\,\text{MPa}$ and $\beta_{test} = 6$. This distribution represents the breakage stress of as-received float glass specimens in an R400 coaxial double ring test at a stress rate of $2 \pm 0.4\,\text{MPa/s}$ and with $A_{test} = 0.24\,\text{m}^2$ (95% confidence level). These tests were performed as a basis for DIN 1249-10:1990. A 'design bending strength of damaged specimens' $\sigma_{bB,A_{test},d} = 24.7\,\text{MPa}$ is defined as the 1.2‰ fractile value of the failure strength distribution proposed by Blank (1993). This distribution, characterized by $\theta_{A_{test},\text{limit}} = 32\,\text{MPa}$ and $\beta_{\text{limit}} = 25$, was chosen based on laboratory tests with the same setup as described above but on specimens with artificially induced homogeneous surface damage. The chosen distribution is somewhat more conservative than the actual test results. The partial resistance factor is defined as $\gamma_M = \sigma_{bB,A_{test},k}/\sigma_{bB,A_{test},d} \approx 1.80$.

This approach is reused in unaltered form for the European draft code prEN 13474-1:1999, see Section 3.2.4.

Extension for beams

Equation (3.3) is adapted for beams by adapting the coefficients:

$$\sigma_{\max,d} \leq \frac{\sigma_{bB,L_{test},k}}{\alpha_\sigma(q,\sigma_V)_{BZ} \cdot \alpha(L_{red}) \cdot \alpha_{BZ}(t) \cdot \alpha_{BZ}(S_v) \cdot \gamma_{M,E}} + \frac{\sigma_{Vk}}{\gamma_{M,V}} \quad (3.6)$$

$$\alpha_\sigma(q,\sigma_V)_{BZ} = \left[\frac{1}{L_{red}} \int_{L_{red}} \left(\frac{\sigma_1(l)}{\sigma_{\max,d}}\right)^\beta dl\right]^{1/\beta} \qquad \alpha(L_{red}) = \left(\frac{L_{red}}{L_{test}}\right)^{1/\beta} \qquad \alpha(t) \approx 3.7 \quad (3.7)$$

$\sigma_{bB,L_{test},k}$ is the characteristic bending strength (5% fractile) of beam specimens with decompressed length L_{test} (= 0.46 m). $\sigma_1(l)$ is the major principal stress at location l. $\alpha(L_{red})$ accounts for the length of a beam's decompressed edge. Sedlacek et al. (1999) recommends the use of $\beta = 5$ for polished and $\beta = 12.5$ for unpolished edges. The values were determined from very small samples (11 and 13 specimens respectively). $\gamma_{M,E} \approx 1.40$ is proposed for $\beta = 12.5$. $\alpha_\sigma(q,\sigma_V)_{BZ}$ equals 1.0 for a uniform stress distribution, 0.94 for a parabolic and 0.86 for a triangular one. $\alpha_{BZ}(S_v)$ is equal to $\alpha(S_v)$.

3.2.4 European draft standard prEN 13474

The design method of prEN 13474[9] (prEN 13474-1:1999; prEN 13474-2:2000) is based on the DELR design method, but contains influences from the methods of Shen and Siebert (see Sections 3.2.5 and 3.2.6). The draft standard faced stiff opposition and is still under revision at the time of writing. The influence of the stress distribution on the glass surface is accounted for on the action side of the verification equation, the residual surface stress on the resistance side. The structural safety verification format compares an *effective stress* σ_{eff} with an *allowable effective stress for design* $f_{g,d}$:

$$\sigma_{\text{eff},d} \leq f_{g,d} \quad (3.8)$$

The effective stress $\sigma_{\text{eff},d}$[10] has to be determined for the most unfavourable action combination as:

$$\sigma_{\text{eff},d} = \left[\frac{1}{A}\int_A (\sigma_1(x,y))^\beta \, dx dy\right]^{1/\beta} \quad (3.9)$$

---

[9] **Important**: This standard is currently under revision by the committees CEN/TC 250 ('Structural Eurocodes') and CEN/TC 129 ('Glass in Buildings'). At the time of writing, the non-public working papers differ considerably from the published draft standards prEN 13474-1:1999 and prEN 13474-2:2000.

[10] prEN 13474 does not use the index d, even when referring to the design level. The index is added here for clarity.

3.2. DESIGN METHODS

A is the *total* surface area of the glass pane and $\sigma_1(x,y)$ is the major principal stress *due to actions* at the point (x,y) on the surface. This means that the effective stress is defined *independently* from the residual stress and that decompression of the whole surface is assumed. Using the coefficient for annealed glass $\alpha_\sigma(p)$ from the DELR design method, it is $\sigma_{\text{eff},d} = \sigma_{\text{max},d} \cdot \alpha_\sigma(q)$. β is the shape parameter of the Weibull distribution of the breakage stress. For common geometries and support conditions, prEN 13474-2:2000 provides tables and equations to determine $\sigma_{\text{eff},d}$ in function of the applied load q and the plate dimensions without actually having to solve Equation (3.9).

The *allowable effective stress* is defined as:

$$f_{g,d} = \left(k_{\text{mod}} \frac{f_{g,k}}{\gamma_M \cdot k_A} + \frac{f_{b,k} - f_{g,k}}{\gamma_V} \right) \cdot \gamma_n \tag{3.10}$$

$f_{b,k}$ characteristic value of the fracture strength (5% fractile);
       $f_{b,k} = f_{g,k}$ for ANG, 70 MPa for HSG and 120 MPa for FTG

$f_{g,k}$ characteristic value of the inherent strength (5% fractile);
       $f_{g,k} = 45$ MPa for soda lime silica and borosilicate glass

$f_{b,k} - f_{g,k}$ the contribution of residual stress to the failure strength; 0 for annealed glass

γ_V partial factor for the residual stress due to tempering (= 2.3 for SLS glass)

γ_M partial factor for the inherent strength (= 1.8 for SLS glass)

γ_n national partial factor (mostly = 1.0)

k_A coefficient to account for the surface area, defined independently from the residual stress as $k_A = A^{0.04}$ (from Equation (3.5) with mit $A_{\text{test}} = 1\,\text{m}^2$ and $\beta = 25$)

k_{mod} modification factor to account for load duration, load combination and environmental conditions; k_{mod} is given for the following dominant actions: short duration (wind): 0.72, medium duration (snow, climate loads for IGUs): 0.36, permanent loads (self weight, altitude for IGUs): 0.27

In comparison to the DELR design method, prEN 13474 contains the following modifications:

- The factor to account for the influence of the stress distribution on the surface is defined independently from the residual stress.
- k_{mod} replaces $\alpha(t)$ and $\alpha(S_v)$.
- k_A replaces $\alpha(A_{\text{red}})$, but is based on the total instead of the decompressed surface area, which makes it independent of the residual stress. Additionally, it is defined with respect to a reference surface of $1\,\text{m}^2$ instead of $0.24\,\text{m}^2$. Surprisingly, it is used together with the unaltered characteristic strength value which is based on $A_0 = 0.24\,\text{m}^2$.

As a result of these modifications, the partial factors are not directly comparable. The replacement of $\alpha(t)$ and $\alpha(S_v)$ by k_{mod} is very similar to Shen's concept, but k_{mod} is not identical to η_D. Instead of explicitly accounting for the relative magnitudes of the different loads of a load combination, the few tabulated k_{mod} values include implicit assumptions. Appendix E of prEN 13474-2:2000 proposes a step-by-step procedure for design, using predefined load combinations.

Excursus: Maintaining consistency is far from trivial

The characteristic value of the inherent strength of float glass is said to be $f_{g,d} = 45$ MPa. This value was originally defined in DIN 1249-10:1990, based on coaxial double ring tests on new annealed glass specimens with a surface area of $A_{\text{test}} = 0.24\,\text{m}^2$ (cf. Section 3.1.2). A two-parameter Weibull distribution was fitted to the measured failure stresses. The Weibull parameters obtained were $\theta_{A_{\text{test}}} = 74$ MPa and $\beta = 6$ (at 0.95 confidence level). The characteristic value is defined as the 5% fractile value of this distribution, which gives the 45 MPa mentioned above. To account for the size effect, $f_{g,d}$ is divided by a size factor k_A defined as

$$k_A = A^{0.04} \tag{3.11}$$

with A being the total surface area of the glass plate. As discussed in Section 3.3.6, the actual size factor based on Weibull statistics is:

$$k_{A,Wb} = (A/A_{test})^{1/\beta} \qquad (3.12)$$

A and A_{test} are the decompressed surface areas of the element to be designed and the specimen used to determine the characteristic strength. This means that:
- A characteristic resistance determined from a distribution with $\beta = 6$ is combined with a correction factor based on $\beta = 25$ (exponent $0.04 = 1/25$).
- The size factor k_A becomes 1 for $A = 1\,\text{m}^2$. This means that the surface area in the tests leading to $f_{g,d}$ is assumed to be approximately four times bigger than it actually was. For $\beta = 25$, the quantitative effect of this is relatively small. Using the real test surface $A_{test} = 0.24\,\text{m}^2$, it is $k_{A,Wb}(A = 1\,\text{m}^2) = 1.059$ (difference of 'only' 6%). For $\beta = 6$, however, it is $k_{A,Wb}(A = 1\,\text{m}^2) = 1.269$ (difference of 27%).

3.2.5 Design method of Shen

Shen presented this design method in Shen (1997). In Wörner et al. (2001), it was adapted to the format of EN 1990:2002. It is mainly a considerable simplification of the DELR design method, with one important exception: The concept to account for residual stresses is taken from the Canadian Standard CAN/CGSB 12.20-M89. Only two coefficients, both on the resistance side of the verification equation, are used and very simple tables are proposed for their values. The residual surface stress of tempered glass is accounted for indirectly by these coefficients. The design method is confined to glass panes made of annealed or fully tempered glass with continuous lateral support along all four edges that are exposed to lateral load. An application to structural elements such as beams or columns is not immediately possible. The structural safety verification format is:

$$\sigma_{max,d} \leq \sigma_k \cdot \frac{\eta_F \cdot \eta_D}{\gamma_R} \qquad (3.13)$$

$\sigma_{max,d}$ design value of the maximum principal stress

σ_k characteristic value of the bending strength determined in R400 coaxial double ring tests (cf. Section 3.1.2).

η_F coefficient to account for surface area stress distribution

η_D coefficient to account for load duration

γ_R partial factor for the resistance

The verification has to be done separately for every different load duration. The factor η_F for the surface area and the stress distribution is defined in a simplistic way, see Table 3.4. The load duration factor[1] η_D is a function of the glass type and is given in Table 3.5. To derive these values, the surface condition and the environmental conditions in structural applications have been assumed to be identical to those in the bending strength laboratory tests. With this assumption, it is

$$\eta_D = \frac{\sigma_D}{\sigma_R} = \left[\frac{t_R}{t_D} \cdot \frac{1}{n+1} \right]^{\frac{1}{n}} \qquad (3.14)$$

σ_D equivalent strength

σ_R bending strength found in laboratory tests (cf. Footnote 11)

t_R test duration[11]

t_D load duration

---

[11] The bending strength values from DIN 1249-10:1990, which refer to tests with a stress rate of $2\,\text{MPa/s}$, are used: $t_{R,ANG} = 45\,\text{MPa} / 2\,\text{MPa/s} = 22.5\,\text{s}$, $t_{R,FTG} = 120\,\text{MPa} / 2\,\text{MPa/s} = 60\,\text{s}$.

3.2. DESIGN METHODS

n crack velocity parameter; $n_{ANG} \approx 17$ (annealed glass), $n_{FTG} = 70$ (fully tempered glass)

The value $n_{FTG} = 70$ is taken from CAN/CGSB 12.20-M89 without further discussion. It has two drawbacks: (a) Increasing a crack velocity parameter is unrelated to the actual physical phenomena governing the resistance of heat treated glass (cf. Section 4.2). (b) While the value of 70 is indeed given in CAN/CGSB 12.20-M89, it does not relate to the crack velocity parameter n in this standard, despite the use of the symbol n. It is actually the value of a parameter combining n with a constant for the relationship between lateral load and stress in rectangular plates (see Section 3.2.9).

For a combination of loads of different duration, η_D has to be calculated individually. Shen (1997) makes proposals on how this should be done for the combination of snow and dead load as well as for snow and wind.

The choice of the partial factor γ_R for the resistance depends on the target reliability level and the scatter of the bending strength data. Based on the assumption that the bending strength's coefficient of variation is 0.1, Shen (1997) proposes $\gamma_R \approx 1.25$ for buildings of medium importance.[12] Wörner et al. (2001) provides no value for γ_R.

Table 3.4:
Factor η_F for Shen's design method (Wörner et al., 2001).

	$A = 0.5 - 4.0\,m^2$	$A = 4 - 10\,m^2$
ANG	1.0	0.9
FTG	1.0	1.0

Table 3.5:
Factor η_D for Shen's design method (Shen, 1997).

	Dead load (50 yr)	Snow (30 days)	Wind (10 min)
ANG	0.27	0.45	0.69
FTG	0.74	0.83	1.00

3.2.6 Design method of Siebert

This design method was proposed by *Siebert* in Siebert (1999). The major modifications with respect to the aforementioned methods are as follows:

- An approach to account for the influence of biaxial stress fields is proposed.
- The residual stress is considered as an action.

The structural safety verification format is

$$\sigma_{ges,d,max} \cdot f_A \cdot f_\sigma \cdot f_{tS} \leq \frac{\theta}{f_P} \qquad (3.15)$$

$\sigma_{ges,d,max}$ maximum principal surface stress; $\sigma_{ges,d,max} = \sigma_{d,max} + \sigma_E$

$\sigma_{d,max}$ maximum principal stress due to actions

σ_E residual surface stress (compression $\Rightarrow$ negative sign)

f_A coefficient to account for the different surface areas of test specimen and actual structural element

f_σ coefficient to account for the different stress distributions in the test specimen and the actual structural element

f_{tS} coefficient to account for load duration and relative magnitudes of different loads

θ scale parameter of the Weibull distribution fitted to experimental bending strength data (has the dimension of a stress)

f_P factor to account for the target failure probability

---

[12] To find this γ_R, the resistance is assumed to follow a log-normal distribution. It can be seen in Section 5.1 that this leads to a consistency problem with the size effect. Shen (1997) does not comment on this issue.

The *stress due to action* is calculated as in the DELR design method. The residual stress $\gamma_V \sigma_V$, however, is considered as an action. Its partial factor cannot be defined on a firm scientific basis due to the lack of data. Siebert proposes $\gamma_V = 1.25$, which conceptually means putting the residual stress back to the resistance side. For a favourable action, the partial factor should rather be < 1.0, which is $1/\gamma_V$.

To obtain *resistance data*, Siebert recommends standard R400 coaxial double ring tests according to EN 1288-2:2000 on specimens with artificially induced homogeneous surface damage and data analysis according to DIN 55303-7:1996, see Section 3.1.2. If tests are performed on heat treated glass, the residual stress has to be deduced from the breaking stress. Siebert proposes, however, to use annealed glass for testing because (a) measurement of the residual stress is imprecise and (b) defects caused by a given method of artificial damaging are more severe in annealed than in heat strengthened or fully tempered glass.

To account for a non-homogeneous stress distribution within the element, the use of a so-called *effective area* $A_{N,ef}$ is proposed:

$$A_{N,ef} = \int_A \left(\frac{\chi \cdot \sigma_{ges,d}(x,y)}{\sigma_{ges,d,max}} \right)^\beta dA \qquad (3.16)$$

$\sigma_{ges,d}(x,y)$ first principal design stress at the point (x,y) on the surface; this refers to the crack opening stress $\Rightarrow \sigma_{ges,d}(x,y) \geq 0$.

$\sigma_{ges,d,max}$ maximum first principal design stress on the surface

A surface area of the glass pane

χ correction factor for the ration of major and minor principal stress; (conservative assumption: 1.0; for a uniaxial stress field, $\chi \approx 0.83$ is proposed)

Using $A_{N,ef}$, a coefficient to account for the difference in surface areas of test specimens and actual structural elements is defined as:

$$f_{A\sigma} = \left(\frac{A_{N,ef}}{A_{L,ef}} \right)^{1/\beta} \qquad (3.17)$$

$A_{L,ef}$ is the effective area of the test specimen. To simplify design tables, it is proposed to split $f_{A\sigma}$ into two factors as follows:

$$f_A = \left(\frac{A}{A_{L,ef}} \right)^{1/\beta} \qquad f_\sigma = \left(\frac{A_{N,ef}}{A} \right)^{1/\beta} = \frac{\sigma_{ges,d,ef}}{\sigma_{ges,d,max}} \qquad (3.18)$$

The effective principal stress $\sigma_{ges,d,ef}$ is defined such that $A \cdot \sigma_{ges,d,ef}^\beta = A_{eff} \cdot \sigma_{ges,d,max}^\beta$. As residual stress is considered as an action, f_σ depends on it. f_A is identical to $\alpha(A)$ in the DELR design method.

The load duration, the relative magnitude of different loads in a load combination and the environmental conditions are accounted for by the factor f_{tS}, which is the product of the factors $\alpha(t)$ and $\alpha(S_v)$ from Güsgen (1998) (Siebert (1999) uses identical assumptions and equations). An additional factor, f_P, enables a target probability of failure G_a to be chosen. It is defined as

$$f_P = \left[\ln \left(\frac{1}{1 - G_a} \right) \right]^{-1/\beta} \qquad (3.19)$$

Siebert (1999) uses the failure probabilities proposed by Blank (1993) and Güsgen (1998). According to these, it is e.g. $1.5 \cdot 10^{-3}$ for structures of medium importance, which gives $f_P = 1.30$ when using $\beta = 25$ from Blank (1993).

In comparison with that used in the DELR design method, Siebert's partial factor for the residual compressive surface stress is clearly less conservative. For annealed glass, both methods give basically identical results despite the different partial factors: $\sigma_{Rd} = \sigma_{bB,A_{test},k}/\gamma_M = 45\,\text{MPa}/1.8 \approx \theta/f_P =$

3.2. DESIGN METHODS

32 MPa/1.3 ≈ 25 MPa. The reason for the different factors is that the resistance is based on the Weibull scale parameter (θ) in Siebert's method and on the characteristic strength ($\sigma_{bB,A_{test},k}$) in the DELR design method.

3.2.7 Glass failure prediction model (GFPM)

The glass failure prediction model (GFPM) presented in Beason (1980) and Beason and Morgan (1984) is directly based on the statistical theory of failure for brittle materials advanced by Weibull (1939). According to Weibull, the failure probability of a brittle material can be represented as

$$P_f = 1 - e^{-B} \tag{3.20}$$

where B reflects the risk of failure as a function of all relevant aspects, in particular the surface condition and the stress distribution. For general cases, the GFPM proposes the risk function

$$B = \tilde{k} \int_A \left[\tilde{c}(x,y) \sigma_{eq,max}(q,x,y) \right]^{\tilde{m}} dA \tag{3.21}$$

in which $\tilde{c}(x,y)$ is the 'biaxial stress correction factor' (a function of the minor to major principal stress ratio), A the surface area and $\sigma_{eq,max}(q,x,y) = \sigma(q,x,y)(t_d/60)^{1/16}$ the maximum equivalent principal stress as a function of the lateral load q and the point on the plate surface (x,y). $\tilde{m}$ and $\tilde{k}$ are the so-called 'surface flaw parameters'.[13] Based on this, the following expression is introduced for rectangular glass plates exposed to uniform lateral loads of constant duration:

$$B = \tilde{k}(ab)^{1-\tilde{m}} \left(Eh^2 \right)^{\tilde{m}} \left(\frac{t_d}{60} \right)^{\tilde{m}/16} \tilde{R}\left(\tilde{m}, \tilde{q}, \frac{a}{b} \right) \tag{3.22}$$

a and b are the rectangular dimensions of the plate ($a > b$), h is the effective thickness, t_d is the load duration in seconds and E is Young's modulus (71.7 GPa in the GFPM). The non-dimensional function

$$\tilde{R}\left(\tilde{m}, \tilde{q}, a/b \right) = \frac{1}{ab} \int_A \left[\tilde{c}(x,y) \tilde{\sigma}_{max}(\tilde{q},x,y) \right]^{\tilde{m}} dA \tag{3.23}$$

depends on the surface flaw parameter $\tilde{m}$ and the distribution of the non-dimensionalized stress on the surface. $\tilde{q} = q(ab)^2/(Eh^4)$ is the non-dimensionalized load and $\tilde{\sigma} = \sigma(q,x,y)ab/(Eh^2)$ is the non-dimensionalized stress.

The surface flaw parameters $\tilde{m}$ and $\tilde{k}$ cannot be measured directly. They are determined from constant load rate tests on rectangular glass plates using a rather complex iterative procedure. In order to establish the stress/time relationship at the location of the flaw that caused failure, the failure origin has to be determined visually. From this relationship, the 60 s equivalent failure stress and the corresponding 60 s equivalent failure load is calculated. Then, a set of risk factors, $\tilde{R}\left(\tilde{m}, \tilde{q}, a/b \right)$, corresponding to each equivalent failure load is calculated for a wide range of assumed $\tilde{m}$. The best value of $\tilde{m}$ is determined by choosing the one which results in a coefficient of variation of the risk factor closest to 1.0 ($\Rightarrow$ mean = standard deviation). $\tilde{k}$ can then be calculated using the plate's geometry and the mean of the set of $\tilde{R}\left(\tilde{m}, \tilde{q}, a/b \right)$ for the best $\tilde{m}$. Both its magnitude and its units are dependent on $\tilde{m}$.

Some minor improvements of the GFPM and its implementation in ASTM E 1300 are presented in Beason et al. (1998) and integrated into recent versions of the standard.

3.2.8 American National Standard ASTM E 1300

The American National Standard 'Standard Practice for Determining Load Resistance of Glass in Buildings' ASTM E 1300-04 provides extensive charts to determine the required thickness of glass

---

[13] The tildes are not used in the source. They are required in the present document to avoid confusion in subsequent chapters.

plates. It is based on the glass failure prediction model by Beason & Morgan (see Section 3.2.7) and on the finite difference stress and deflection analysis by Vallabhan (Vallabhan, 1983). Resistance is defined using a target failure probability of 8‰. ASTM E 1300 applies to vertical and sloped glazing in buildings exposed to a uniform lateral load and made of monolithic, laminated, or insulating glass elements of rectangular shape with continuous lateral support along one, two, three or four edges. The specified design loads may consist of wind load, snow load and self-weight with a total combined magnitude less than or equal to 10 kPa. The standard does *not* apply to other applications such as balustrades, glass floor panels and structural glass members or to any form of wired, patterned, etched, sandblasted, drilled, notched or grooved glass or to any glass with surface and edge treatments that alter the glass strength. The verification format is

$$q \leq \text{LR} = \text{NFL} \cdot \text{GTF} \tag{3.24}$$

with q being the uniform lateral load, LR the 'load resistance', NFL the 'non-factored load' (based on a 3 s load duration) and GTF the so-called 'glass type factor' (load-duration dependent, see below).

The important difference with respect to European design methods is that this verification format is based on *loads* and not on *stresses*. Furthermore, it does not use any partial factors. The NFL is determined from charts given for various geometries, support conditions, glass thicknesses and for monolithic as well as laminated glass. The GTF combines glass type and load duration effects and is given for single panes (Table 3.6) as well as for insulating glass units.

Glass type	Short duration load	Long duration load
ANG	1.0	0.5
HSG	2.0	1.3
FTG	4.0	3.0

Table 3.6: Glass type factors (GTF) for a single pane of monolithic or laminated glass.

All charts and values are calculated using the GFPM with $\tilde{m} = 7$, $\tilde{k} = 2.86 \cdot 10^{-53} \text{N}^{-7} \text{m}^{12}$, a Young's modulus of $E = 71.7 \text{ GPa}$ and the effective (not the nominal) glass thickness (ASTM E 1300-04; Beason et al., 1998). The non-factored load charts incorporate the viscoelastic model for the plastic interlayer from Bennison et al. (1999). This model claims to describe accurately the evolution of the polymer shear modulus at 50 °C. At this temperature and for a load duration of 3 s (the reference in the standard), the PVB interlayer is characterized with an effective Young's modulus of 1.5 MPa. This value is meant to be a lower bound for commercially available PVB interlayers.

For independent stress analyses required in the case of special shapes or loads not covered in the standard, allowable surface stresses for a 3 s duration load are given, see Table 3.7. The values for edges are taken from Walker and Muir (1984). It is claimed for the allowable 3 s stress and $P_f < 0.05$ in annealed glass away from the edges, that the following equation should give conservative values:

$$\sigma_{\text{allowable}} = \left(\frac{P_f}{\tilde{k}(d/3)^{7/n} A} \right)^{1/7} \tag{3.25}$$

The constant 7 in Equation (3.25) is the parameter $\tilde{m}$ and 3 is the reference time period in seconds. For $P_f = 0.008$, $d = 3 \text{ s}$ and $A = 1 \text{ m}^2$, Equation (3.25) yields 16.1 MPa, which is indeed very conservative with respect to the value of 23.3 MPa given in Table 3.7.

	annealed glass	heat strength- ened glass	fully tempered glass
away from the edges	23.3	46.6	93.1
clean cut edges	16.6	n/a	n/a
seamed edges	18.3	36.5	73.0
polished edges	20.0	36.5	73.0

Table 3.7: Allowable surface stress (MPa) for a 3 s duration load according to ASTM E 1300-04.

3.2. DESIGN METHODS

To be able to compare the allowable stresses in Table 3.7 to those from Table 3.3, they must be converted to the same reference time period ($\sigma_{60s} = \sigma_{3s}(3/60)^{1/16} = \sigma_{3s} \cdot 0.829$). For annealed glass, very similar values are obtained. For fully tempered glass, the allowable stress is clearly higher according to ASTM E 1300-04 ($\sigma_{60s} = 77.2\,\text{MPa}$) than according to TRLV 1998 ($\sigma_{60s} = 50\,\text{MPa}$).

The 3 s duration load that represents the combined effects of I loads of different duration (all normal to the glass surface) is determined using[14]

$$q_3 = \sum_{i=1}^{I} q_i \left[\frac{d_i}{3}\right]^{1/n} \qquad (3.26)$$

where q_3 is the magnitude of the 3 s duration uniform load and q_i the magnitude of the load having duration d_i. For annealed glass, it is $n = 16$.

3.2.9 Canadian National Standard CAN/CGSB 12.20

The Canadian National Standard 'Structural Design of Glass for Buildings' CAN/CGSB 12.20-M89 deals with soda lime silica glass panes exposed to uniform lateral load. Like the American National Standard, it is based on the GFPM (see Section 3.2.7) and a target failure probability of $P_f = 0.008$ for the resistance. It is important to notice that in contrast to ASTM E 1300-04, which uses a 3 s reference duration for the resistance, CAN/CGSB 12.20-M89 is based on a 60 s reference duration. This is due to the fact that the Canadian Standard, published in 1989, is based on ASTM E 1300-94 while the 3 s reference duration was only introduced in ASTM E 1300-03.

Standard cases

For standard cases, the verification format is as follows:

$$E_d \leq R_d \qquad (3.27)$$

E_d combination of all actions (design level = including partial factors)

R_d resistance of the pane (design level = including partial factors)

The action term is:

$$E_d = \alpha_D D + \gamma \cdot \psi \cdot (\alpha_L L + \alpha_Q Q + \alpha_T T) \qquad (3.28)$$

D dead loads (self weight, invariant hydrostatic pressure)

L live load (snow, rain, use and occupancy, variable hydrostatic pressure)

Q live loads (wind, stack effect, earthquake, climatic and altitude load for IGUs)

T effects of temperature differences except those included in Q

α_x partial factors: $\alpha_D = 1.25$ (unfavourable) or 0.85 (favourable); $\alpha_L = \alpha_Q = 1.50$; $\alpha_T = 1.25$

γ importance factor: $\gamma = 1.0$ (in general), $\gamma \geq 0.8$ (farm buildings having low human occupancy or buildings for which collapse is not likely to cause serious consequences)

ψ load combination factor: $\psi = 1.0$ (when only one of L, Q and T acts), $\psi = 0.7$ (when two of L, Q and T act), $\psi = 0.6$ (when all of L, Q and T act). The combination with the most unfavourable effect has to be determined.

---

[14] This equation is incorrect, see Section 5.3.2.

The resistance term is:

$$R = c_1 \cdot c_2 \cdot c_3 \cdot c_4 \cdot R_{ref} \qquad (3.29)$$

c_1 glass type factor: 1.0 (flat glass, laminated glass), 0.5 (sand blasted, etched or wired glass)

c_2 heat treatment factor: 1.0 (annealed glass), 2.0 (heat strengthened glass), 4.0 (fully tempered glass)

c_3 load duration coefficient

load type	approx. equiv. duration	ANG	HSG	FTG
wind and earthquake	1 min	1.0	1.0	1.0
sustained (snow, ponding)	1 week to 1 month	0.5	0.7	0.9
continuous (dead load, hydr. pressure)	1 year to 10 years	0.4	0.6	0.8

c_4 load sharing coefficient (for insulating glass units): 1.0 (monolithic glass), 1.7 and 2.0 (double-glazed and triple-glazed sealed insulating glass units with similar glass types and thicknesses)

R_{ref} reference factored resistance of glass (the standard gives tabulated values) (factored resistance of annealed glass loaded to failure under a constant load in 60 s; the values given are based on the minimum allowable (not the nominal) thickness and an expected failure probability of 0.8%)

Laminated glass may be considered as monolithic glass if the load duration is < 1 minute and the temperature < 70 °C or if the load duration is < 1 week and the temperature < 20 °C. For any other condition, laminated glass has to be considered as layered glass (no composite action may be assumed).

Special cases

For non-standard applications that are not covered by the tables and factors, some more general indications are given. They allow to get more insight into the model that the tabulated values are based on. The area effect is accounted for by

$$R_A = R_{ref} \cdot A^{(-1/\tilde{m})} \qquad (3.30)$$

where A is the area of the pane in m^2 and $\tilde{m}$ 'varies from about 5 to 7'. The load duration effect is accounted for by

$$R_t = R_{ref} \cdot t^{(-1/\tilde{n})} \qquad (3.31)$$

with t being the load duration in minutes (!) and $\tilde{n}$ being 15 for ANG, 30 HSG and 70 for FTG.

It is crucial to notice that $\tilde{n}$ is *not* equal to the exponential crack velocity parameter n although the letter 'n' is used in CAN/CGSB 12.20-M89. The tilde has therefore been added here to avoid confusion. Based on Johar (1981, 1982), the standard assumes that $\sigma \propto R^c$, σ being the 'stress in fracture origin areas', R the uniform lateral load and c a constant < 1. It is $\tilde{n} = cn$, which means that Equation (3.31) is in fact a combination of Brown's integral (see Section 4.2.4) with the proportionality between the stress and q^c found for rectangular plates (q is the uniform lateral load). As this proportionality and the value of c are included in $\tilde{n}$, the tables and equations in the Canadian Standard should not be applied to other geometries, boundary conditions or loading conditions. The value of c is not directly given in the standard, but as n is said to be 16 (called d in the terminology of the standard) and $\tilde{n} = 15$, it should be 15/16.

For general cases, CAN/CGSB 12.20-M89 recommends to limit stresses to 25 MPa away from the edges of plates and to 20 MPa on clean-cut edges. These values have to be corrected by the factor for the area effect and most probably also for the load duration, although the latter is not mentioned explicitly.

3.3. ANALYSIS OF CURRENT GLASS DESIGN METHODS

Note

The n versus $\bar{n}$ issue gives rise to a certain number of problems and misunderstandings. This has already been seen when discussing Shen's design method (Section 3.2.5), but it also affects the Canadian standard itself. In Appendix B of the standard, the one-minute reference resistance R_{ref} is said to be (R_{f} is the failure load at the time of failure t_{f} in minutes):

$$R_{\text{ref}} = R_{\text{f}} \left(\frac{t_{\text{f}}}{\bar{n}+1} \right)^{1/\bar{n}} \tag{3.32}$$

Using the equations in Section 4.2.5 and $\sigma \propto R^c$, it is seen, however, that it should be[15]:

$$R_{\text{ref}} = R_{\text{f}} \left(\frac{t_{\text{f}}}{n+1} \right)^{1/\bar{n}} \tag{3.33}$$

3.3 Analysis of current glass design methods

3.3.1 Introduction

After taking a close look at specific testing procedures and design methods in Sections 3.1 and 3.2, it is now pertinent to

- establish an overview;
- identify the main weaknesses and limits of applicability of current glass design methods;
- define the topics that need further investigation.

The preceding sections have shown that most design methods are actually variations, extensions or simplifications of others. There are two groups: *European design methods*, which are based on the DELR design method, and *North American design methods*, which are based on the GFPM.

3.3.2 Main concepts

In Table 3.8, European and North American design methods are compared with respect to the main concepts that they are based on. It can be seen that the two approaches are not directly comparable because of conceptual incompatibilities. Data from experiments designed for one method cannot directly be used in conjunction with the other method.

It would be advantageous to develop a general glass model that is able to supersede all current model variants and specialized models. Furthermore, such a model should be based on a set of parameters that can be determined from various experimental setups. Such a model could enable the development, in the medium term, of a versatile design method that caters for all element geometries, support conditions and loading conditions.

3.3.3 Obtaining design parameters from experiments

Laboratory testing involves extremely short load durations compared to the typical design life of glass structures. As the material's behaviour is strongly time-dependent because of subcritical crack growth, extreme caution must be exercised when extrapolating beyond the laboratory data range to long service lifetimes. Although it would be safer to determine design parameters on specimens in their in-service condition and environment and with load durations similar to typical service lifetimes, this is hardly ever possible. Such tests would be too costly and time-consuming.

At present, experimental tests for the determination of design parameters are performed at ambient conditions. The parameters meant to represent the surface condition or a 'material strength' ($\bar{k}$ and $\tilde{m}$ in the North American design methods, θ_A and β in the European design methods) are, therefore,

---

[15] Though this has already been pointed out in Fischer-Cripps and Collins (1995), the standard has not yet been revised at the time of writing (2006).

Table 3.8: Comparison of European and North American design methods.

	European design methods	North American design methods
Testing procedure to obtain strength data	Constant stress rate coaxial double ring tests ($\dot{\sigma}_{test} = 2 \pm 0.4\,\text{MPa/s}$) on specimens with $A_{test} = 0.24\,\text{m}^2$ (cf. Section 3.1.2).	Rectangular glass plates exposed to uniform lateral load.
Surface condition of strength test specimens	Artificially induced homogeneous surface damage.[*]	Weathered windows glass.
Design 'strength' definition	A single value is used, the 'characteristic value of the inherent strength'[†]. It is defined as the 5% fractile value (at 0.95 confidence level) of the failure stress measured in the experiments.	Two interdependent parameters called 'surface flaw parameters' $\widetilde{m}$ and $\widetilde{k}$ are used. Their determination from experimental data is based on the stress history at the visually determined failure origin and a rather complex iterative procedure.
Subcritical crack growth	Taken into account by load duration factors that depend on the loading only. The factors are based on the empirical relationship $v = S \cdot K_I^n$, (cf. Section 4.1).	Only one crack velocity parameter is used explicitly. It is equivalent to the parameter n in European methods and assumed to be 16.
Extrapolation from experiments to in-service conditions	Some but not all differences between laboratory conditions and actual in-service conditions are accounted for by correction coefficients. Details vary between methods, see Section 3.2.	Graphs are provided for many common cases (in terms of geometry and support conditions). They give uniform lateral loads that a given glass pane can withstand for a reference time period.
Taking the glass type into account	Mostly by adding the absolute value of the residual surface stress (multiplied by a 'safety factor') to the allowable tensile stress of float glass.	By multiplying the load resistance of a float glass element by some load duration-dependent glass type factor.[‡]

[*] The generally used parameter set does, however, not directly reflect test data, see Section 3.2.3.
[†] In contrast to usual characteristic resistance values, this one is not a 'real' material parameter. It depends on the geometry, the surface condition, the environment and the loading of the specimens. The term 'characteristic value' is therefore somewhat misleading, which is why it is put in inverted commas.
[‡] In CAN/CGSB 12.20-M89, the glass type factor (called 'heat treatment factor') does not depend on the load duration. The load duration factor, however, is glass type dependent, which comes to the same thing.

inevitably dependent on the surface condition *and* on crack growth behaviour. This is problematic for at least two reasons:

- The parameters combine unrelated physical aspects, namely the effects of surface condition and subcritical crack growth, within a single value.
- If crack growth during the tests is low, high failure strengths will be measured and vice versa. This means that if the crack velocity is faster under in-service conditions than during the tests, design based on experimentally determined strength values is *unsafe*.

3.3.4 Load duration effects

Time-dependent effects related to loads are commonly referred to as 'load duration effects' or 'duration-of-load effects'. Strictly speaking, the term is not very accurate because it implies constant loads. In the more general case of time-variant actions, 'action history effects' would be more appropriate. As 'load duration effect' represents commonly accepted terminology and is widely used in academic publications, the term is nevertheless used in the present document.

3.3. ANALYSIS OF CURRENT GLASS DESIGN METHODS

Resistance

Glass strength is time-dependent because of stress corrosion (see Section 4.1). The design methods presented in Section 3.2 conceal this dependence within coefficients, such that the underlying assumptions are not readily visible. The present section deals only with the choice of parameters; the basis of crack growth modelling will be discussed in Section 4.1.

The 'classic' European crack velocity parameters have been published in Kerkhof et al. (1981). They are based on the ambient condition crack growth data from Richter (1974). He determined those parameters by optically measuring the growth of large through-thickness cracks on the edge of specimens loaded in uniform tension (see Section 3.1.3). On this basis, 'design parameters' for the DELR design method were chosen in Blank (1993). (These design parameters represent substantially higher crack velocities than Richter's measurements, see Section 6.3). The European draft standard prEN 13474 and the design method by Siebert are directly based on the DELR design method and use the same parameters. Shen uses a different approach, see Section 3.2.5.

The GFPM uses, as stated above, only one crack velocity parameter explicitly. It is equivalent to the parameter n in European design methods and assumed to be equal to 16.

In conclusion, current glass design methods assume that the parameters of the crack growth model are well known, stable, deterministic values that are identical in laboratory tests and under in-service conditions. The actual values are based on few experiments and it is uncertain whether these experiments accurately represent in-service conditions. Further investigations as to the reliability and the potential variance of the crack velocity parameters are, therefore, required.

Actions

All modern glass design methods are implicitly or explicitly based on the assumption that crack growth and with it the probability of failure of a crack or an entire glass element can be modelled using the *risk integral*, also known as *Brown's integral*. Section 4.2.4 explains how this integral is obtained by integration of the ordinary differential equation of the crack growth. The main implication of the risk integral is that, if the crack velocity parameter n is constant, two stress histories $\sigma_{(1)}(\tau)\ \tau \in [0, t_1]$ and $\sigma_{(2)}(\tau)\ \tau \in [0, t_2]$ cause the same amount of crack growth if

$$\int_0^{t_1} \sigma_{(1)}^n(\tau)\,d\tau = \int_0^{t_2} \sigma_{(2)}^n(\tau)\,d\tau. \tag{3.34}$$

The risk integral is the basis of any equivalent stress approach (cf. Section 4.2.5). It is, however, at the root of two (related) problems:

- The risk integral is a function of the stress history only. This means that the momentary failure probability at a point in time, which is a function of the risk integral, is actually independent of the momentary load at that time.
- The risk integral approaches zero when the loading time approaches zero. For very short loading times (or very slow subcritical crack growth), the material resistance obtained from an equivalent stress based model converges on infinity. This does not make sense. It should converge on the inert strength, which is an upper resistance limit (cf. Section 4.2.2).

These fundamental issues were among the reasons that led to confusion and a general lack of confidence in advanced glass design methods. Together with the high variance in experimentally determined parameters, they are also the main reasons for various researchers to claim that the glass failure prediction model and any Weibull distribution based design approach are fundamentally flawed and unrealistic (e.g. Calderone (2005, 2001); Reid (1991, 2000); see also Section 3.3.7).

How and to what is the load duration effect applied?

All design methods account for the load duration effect by a factor that depends on the load duration and sometimes the residual stress *only*. In European methods, this factor is applied to the allowable maximum or equivalent in-plane principal stress. In GFPM-based methods, it is applied to the allowable lateral load. The limits of applicability of this approach are not discussed in the design methods. In view of all aspects that influence the load duration effect, however, one would expect the load duration factor to depend on:

- the residual stress[16],
- the action history and action combination,
- the subcritical crack growth model,
- the environmental conditions,
- the element's geometry.

3.3.5 Residual stress

Several methods include the effect of residual stresses within the resistance of the glass. It is, however, crucial to distinguish residual stress clearly from inherent strength for several reasons. The most important ones are:

- Only decompressed parts of a glass element's surface are subjected to subcritical crack growth and its consequences. The behaviour of non-decompressed surfaces is not influenced by factors, such as duration of load, which affect the inherent strength only.
- The uncertainties, and consequently the partial factors, are different for residual stress and inherent strength.

Design methods accounting for residual stress explicitly superimpose the residual stress on the inherent strength. This assumes that the inherent strength is not affected by heat treatment. There is evidence showing that the tempering process actually causes a certain amount of 'crack healing' (Bernard, 2001; Hand, 2000); this assumption can thus be considered safe (conservative) for design.

In order to be consistent with modern standard and design methods for building materials other than glass, residual stress should be considered as a *beneficial action*. It is, however, a property of the *material* which means that it is the manufacturer's responsibility to guarantee that the specifications (e. g. a minimal residual stress) are met. From this perspective, residual stresses are part of product and testing standards, while actions would be part of design standards.

3.3.6 Size effect

As a direct consequence of the use of Weibull statistics, the resistance of glass elements depends on their surface areas in all design methods. As only tensile stress can cause glass failure, the size depends not on the total, but on the decompressed, surface area. For given geometry and support conditions, the latter depends in general[17] on the load intensity and is therefore time-variant. Taking this aspect accurately into account is complex (cf. Chapters 4 and 5). European design methods define the size factor based on the total surface area, which makes it load-independent. US and Canadian standards multiply the load resistance of annealed glass elements by a factor. As the entire surface of an annealed glass plate is immediately decompressed on loading, this comes to the same thing. The size effect can be expressed as (cf. Chapter 4)

$$\frac{\sigma(A_1)}{\sigma(A_2)} = \left(\frac{A_2}{A_1}\right)^{1/s} \tag{3.35}$$

where $\sigma(A_1)$ and $\sigma(A_2)$ are the tensile strengths of structural members with surface areas A_1 and A_2 respectively exposed to tensile stress. In European methods, s is the shape parameter β of the tensile fracture strength distribution. In GFPM based methods, s is the surface flaw parameter $\tilde{m}$.

---

[16] As long as the surface is not decompressed, there is no load duration effect at all. Depending on the load intensity, none, all or parts of the surface may be decompressed.

[17] In some special but frequent cases such as annealed glass plates exposed to uniform lateral load, the whole surface is decompressed at all non-zero load intensities.

3.3. ANALYSIS OF CURRENT GLASS DESIGN METHODS

The two parameters $\tilde{m}$ and β do not have the same meaning. It will be seen in Section 5.4 that it is $\beta = \tilde{m}(n+1)/n$ (so typically $\beta = 17/16 \cdot \tilde{m}$). Figure 3.9 shows that the size effect is quite significant for the ASTM E 1300 value of $\tilde{m} = 7$, while it becomes almost negligible (for realistic panel sizes) for the value $\beta = 25$ that is generally used in European design methods.

Further investigations are required to answer the following questions:

- Why does the exponent s differ that much between design methods?
- Is there a feasible way of accurately taking the size effect into account?
- Why do some experiments show little or no relationship between the total surface area and the breakage stress or between the most stressed panel area and the breakage stress (Calderone, 2001)? Is this proof that Weibull statistics are inadequate for glass modelling? What alternatives exist?

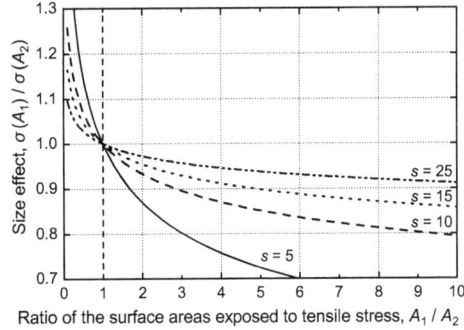

Figure 3.9:
The size effect's dependence on the Weibull shape factor or first surface flaw parameter (using Equation (3.35)).

3.3.7 Weibull statistics

The fit of many experimental data sets to the Weibull distribution is rather poor. Various proposals have been made to address this problem:

- Determine the 'characteristic strength value' by fitting a normal or log-normal distribution instead of a Weibull distribution to the experimental strength data (e. g. Fink (2000); Laufs (2000b)). The size effect, however, which is a direct consequence of the use of Weibull statistics, is still applied in unaltered form.
- Completely abandon statistics and failure probability based design and use the classic theory for the strength of materials, based on zero variation in strength (e. g. Calderone (2001); Jacob and Calderone (2001)).

Both approaches are unsatisfactory and cannot be readily applied with confidence. In-depth investigation of this topic is clearly required. This is done in Section 5.1.

3.3.8 Biaxial stress fields

The GFPM-based standards use the biaxial stress correction factor proposed by the GFPM. It depends on the principal stress ratio (which is, in general, load intensity-dependent) and on the surface flaw parameter $\tilde{m}$. Though not explicitly stated, only the fully-developed principal stress ratio is used for the resistance graphs and the testing procedure. Based on this ratio, a single biaxial stress correction factor for each point on the surface is calculated and assumed to be valid for all load intensities.

European glass design methods generally assume all cracks to be oriented perpendicularly to the major principal stress. This is equivalent to assuming an equibiaxial stress field. While this assumption is conservative (safe) for design, it is *not* conservative when deriving glass strength data from tests. Although it had been observed that the strength values measured in four point bending tests (uniaxial stress field) were systematically higher than those measured in coaxial double ring tests (equibiaxial stress field), this effect was for many years not quantified. Siebert addressed this issue (Siebert, 1999, 2001). He took the view that the biaxial stress correction approach of the GFPM was inappropriate and that an approach more directly based on experiments was required. He developed a clever test setup that allows any principal stress ratio between 0 and 1 to be applied to glass specimens. In this way he was able to establish a purely empirical biaxial stress correction factor. Siebert chose to define this factor in function of the principal stress ratio *only*.

Further investigation is required to:

- Find out whether and how the effect of biaxial stress fields can be accurately modelled using a general and entirely fracture mechanics based approach;
- Assess possible simplifications (including those mentioned above) and their limits of validity.

3.3.9 Consistency and field of application

Current glass analysis and design methods have been developed and improved by numerous researchers and institutions over many years. Not surprisingly, this has led to *consistency problems* as was seen in Section 3.2. They tend to confuse users, make it difficult to understand the models and give rise to doubts about their suitability.

An ideal design method's *field of application* should include more geometries than rectangular glass plates and more loading conditions than uniform lateral loads. In particular concentrated loads, line loads, stability problems and in-plane loads are frequently encountered in structural glass applications. In several glass-specific hazard scenarios, such as vandalism or accidental impact, the danger for a structural element arises not from the load intensity but from surface damage. A straightforward way to analyse these cases should be provided.

3.3.10 Additional information and quality control

The structural efficiency of glass elements suffers from the large uncertainties associated with the material resistance. The large coefficients of variation require high safety margins for design values. A design method should therefore enable its user to take advantage of measures taken to reduce or quantify accurately the coefficient of variation of resistance parameters. Such measures could be:

- Proof loading of elements or parts of the structure after manufacture or even after installation;
- Quality control measures such as direct or indirect (fragmentation count) measurement of the residual stress or visual detection of surface damage;
- Non-structural measures that prevent or limit potential glass surface damage.

3.4 Summary and Conclusions

> ⇨ **European and North American design methods.** Most design methods are actually variations, extensions or simplifications of others. From a conceptual point of view, there are only two groups: European design methods, which are based on the design method of damage equivalent load and resistance, and North American design methods, which are based on the glass failure prediction model.
> ⇨ **Background.** Assumptions, limits of validity and derivations of the models are often unclear. This is mainly due to a historical mix of empirical approaches and physical models and to simplifying assumptions being deeply integrated into the models.
> ⇨ **Design parameters.** Some parameters combine several physical aspects. As a result, the parameters reflect other influences than those they are supposed to and depend on the experimental setup used for their determination.
> ⇨ **Range of validity.** Because of underlying implicit or explicit assumptions, none of the existing models takes all aspects that influence the lifetime of structural glass elements into consideration and none is applicable to general conditions. Common limitations to rectangular plates, uniform lateral loads, constant loads, time-independent stress distributions and the like prevent application to cases such as in-plane or concentrated loads, loads which cause not only time-variant stress levels but also time-variant stress distributions, stability problems, and connections.
> ⇨ **Surface condition.** Through the condition of the specimens used to obtain strength data, an assumption about the surface condition is integrated into current design methods (homogeneous random surface flaw population; the properties vary depending on the method). The condition of the glass surface is not represented by independent design parameters that the user can modify. This is a notable drawback, especially when hazard scenarios that involve surface damage caused by factors such as accidental impact, vandalism or wind-borne debris must be analysed or when data from quality control measures or research are available.
> ⇨ **Consistency.** The models contain inconsistencies. They give unrealistic results for special cases and different models yield differing results.
>
> In order to be able to model and design general structural glass elements with confidence, research is required primarily in the following areas:
> ⇨ An improved lifetime prediction model for structural glass elements should be established. It should provide the following features:
>   - A comprehensive and clear derivation should allow the model, and all assumptions it is based on, to be fully understood.
>   - Accuracy, flexibility and scope of application should be extended with respect to current models.
>   - The model's parameters should each represent only one physical aspect and should be independent of test conditions.
>   - A straightforward way to analyse hazard scenarios that involve severe surface damage should be provided.
>   - A straightforward way to benefit from additional knowledge should be provided. In particular, the model should allow for less conservative design if additional data from quality control measures or research are available.
> ⇨ Fundamental doubts with respect to structural glass design should be discussed and, if possible, resolved.
> ⇨ Better insight should be provided into how testing should be conducted in order to obtain meaningful and safe results.
>
> The remaining chapters of this thesis aim at improving knowledge in these fields.

Chapter

4

Lifetime Prediction Model

In this chapter, a flexible lifetime prediction model for structural glass elements is established based on linear elastic fracture mechanics and the theory of probability. As a prerequisite, the phenomenon of stress corrosion causing subcritical growth of surface flaws is discussed. Second, a model for the time-dependent behaviour of a single flaw is established. The last step is to extend the single-flaw model to a large number of flaws of random depth, location and orientation. This chapter focuses on fundamental concepts. Quantitative aspects are investigated in Chapter 6 and the actual use of the model is discussed in Chapter 8.

4.1 Stress corrosion and subcritical crack growth

4.1.1 Introduction and terminology

In vacuum, the strength of glass is time-independent.[1] In the presence of humidity, however, strength depends on the action history because surface flaws that are exposed to tensile stress grow with time. This section is dedicated to the fundamental aspects and qualitative description of this phenomenon. A quantitative analysis of existing data can be found in Section 6.3.

In the present document, the term '*stress corrosion*' is used to refer to the *chemical phenomenon*. The term '*subcritical crack growth*' is used to refer to the *consequence* of stress corrosion, the *growth of surface flaws*. In academic publications, this distinction is not always made and other terms, such as 'slow crack growth', 'static fatigue', and 'environmental fatigue' are in use. The terms containing 'fatigue' are somewhat misleading because most people, and in particular engineers, associate 'fatigue' with cyclic loading.

4.1.2 The classic stress corrosion theory

The phenomenon of stress corrosion leading to subcritical growth of flaws under static loads is fundamental for the modelling of structural glass elements. It was described in 1899 by *Grenet* (Grenet, 1899). Since that time, subcritical crack growth in glass has been the subject of a great many research projects and there is an impressive number of scientific publications on the subject. A brief overview of some of the most important work on the subject is given below.

Griffith proposed to account for crack growth and brittle fracture based on the surface energy near the crack[1] (Griffith, 1920). *Levengood* determined a relationship between the crack mirror and the

---

[1] Several research projects (see Kurkjian et al. (2003) for an overview) have shown that even in vacuum, the resistance of many glasses is in fact slightly time-dependent. This effect, called '*inert fatigue*', is the result not of stress corrosion but of the increasing thermal energy available as the temperature is increased above absolute zero. Inert fatigue is, however, of no practical relevance for structural engineering applications.

crack depth at failure. He stated that the reciprocal of the breaking stress is approximately a linear function of the depth of deep cracks. He also noted an increase in glass strength when the time period between the surface scratching and the tests was increased, as well as a dependence of the increase in strength on the conditions the glass was exposed to (air, water and various other liquids) (Levengood, 1958).

From 1967, *Wiederhorn* and his staff conducted in-depth investigations on the phenomenon of stress corrosion and subcritical crack growth in glass, also including crack healing (Wiederhorn and Tornsend, 1970; Wiederhorn, 1967; Wiederhorn and Bolz, 1970). Their experiments were conducted using double cantilever beam (DCB) specimens that load a long through-thickness crack in tension. This setup is very similar to what is commonly used for fatigue testing of steel specimens. Crack lengths were measured optically with a precision of ±0.01 mm. Crack velocities of up to 10^{-4} m/s could be measured.

Some of the major quantitative results giving evidence of the fundamental aspects of stress corrosion are reproduced in Section 4.1.6. *Wiederhorn* concluded from these results that stress corrosion is governed by a chemical reaction between water and glass. Based on these experimental results, a lifetime prediction model for glass components was developed by Evans and Wiederhorn (1974)[2]. In this model, the relationship between the crack velocity v and the stress intensity factor K_I (v-K_I relationship) in region I is described by the empirical expression

$$v = S \cdot K_I^n \qquad (4.1)$$

with the crack velocity parameters S and n to be determined from experiments. For n, a value of 15 ± 2 is proposed. Based on measurements by *Richter* (Richter, 1974) to determine the parameters in Equation (4.1) and further research work (Kerkhof (1975, 1977) and others), a comprehensive method for the prediction of the lifetime of glass elements with a known initial crack depth was published by *Kerkhof, Richter* and *Stahn* (Kerkhof et al., 1981). This publication and the crack velocity parameters cited therein (see Table 6.5 on page 82) became the basis for virtually all research and glass design method development in Europe (cf. Chapter 3). Kerkhof, Richter and Stahn also investigated two important questions that arose immediately:

♦ Are the large through-thickness cracks used to determine the crack velocity parameters representative of the behaviour of microscopic flaws (cf. Section 3.1)?

♦ The crack velocity could only be measured in the range of 10^{-5} mm/s $\leq v \leq 10^{-2}$ mm/s. This is clearly above the range that is relevant for design (at 10^{-5} mm/s, a crack grows by 1 mm within 28 hours). Can the v-K_I relationship be extrapolated to much lower crack velocities?

It is claimed in Kerkhof et al. (1981) that Equation (4.1) is a usable approximation at least down to $v \approx 10^{-8}$ mm/s and that the model can represent the behaviour of microscopic flaws at least approximately. The scatter of the data, however, is large. Furthermore, the experimentally measured dependence of the breakage stress on the stress rate was found to differ considerably from the calculations. Even in water, the breakage stress seems to converge on an upper limit for stress rates above a few MPa/s. This indicates that there is not enough time for stress corrosion to take place. This is an important issue discussed in more detail in Chapter 6.

Methods for measuring crack velocity have improved considerably since the early 1980's. Modern technologies such as atomic force microscopy (AFM) allow measurements within a range of 10^{-9} mm/s $\leq v \leq 1$ mm/s (Marlière et al., 2003). The relationship in Equation (4.1) has remained basically uncontested. A comprehensive quantitative overview and comparison of crack growth data is given in Section 6.3.

4.1.3 Relationship between crack velocity and stress intensity

Figure 4.1 shows the simplified, schematic relationship between *crack velocity v* and *stress intensity factor* K_I. For values of K_I close to the fracture toughness K_{Ic} (definition → p. 43) or even above, v is

---

[2] Identical reprint, which is easier to obtain: Evans and Wiederhorn (1984)

4.1. STRESS CORROSION AND SUBCRITICAL CRACK GROWTH

independent of the environment and approaches a characteristic crack propagation speed (about 1500 m/s for soda lime silica glasses) very rapidly. In a narrow region below K_{Ic} (region III), the curve is very steep, v lying between 0.001 m/s and 1 m/s. In inert environments (cf. Section 4.2.2), this curve would extrapolate linearly to lower crack velocity. In normal environments, the behaviour strongly depends on the environmental conditions. As discussed in Section 4.1.2, the empirical relationship

$$v = S \cdot K_I^n \qquad (4.2)$$

provides a good approximation for region I. As the dimension of S depends on n in Equation (4.2), the equivalent formulation

$$v = v_0 \left(\frac{K_I}{K_{Ic}}\right)^n \qquad (4.3)$$

is more useful because its parameters' units are independent ($S = v_0 \cdot K_{Ic}^{-n}$). The *linear crack velocity parameter* v_0 has the dimension length per time, while the *exponential crack velocity parameter* n is dimensionless. When the v-K_I-curve is plotted on logarithmic scales, v_0 represents its position and n its slope. K_{Ic} is a material constant that is known with a high level of precision and confidence (see Section 6.2). Regions I and III are connected by region II, where v is essentially independent of K_I but depends on the amount of humidity in the environment. Below a certain threshold value K_{th} (see Section 4.1.5), no crack growth can be measured.

Figure 4.1:
Idealized v-K_I-relationship.

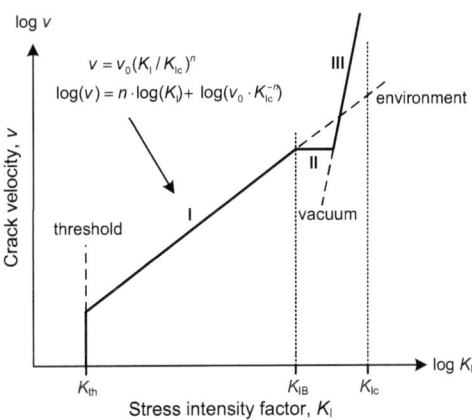

In view of the order of magnitude of glass elements in buildings (mm to m), the typical depth of surface flaws (μm to mm) and the service life generally required, only the range of extremely slow subcritical crack growth, region I, is of interest. The contribution of regions II and III to an element's lifetime is negligible.

Excursus: Power law or exponential function?

The empirical power law in Equation (4.2) is not the only possible approximation for the v-K_I relationship. The exponential functions

$$v = v_i \cdot e^{\beta K_I} \quad \text{or} \quad v = v_i \cdot e^{\beta(K_I - K_{Ic})} \qquad (4.4)$$

are also used in scientific publications. In practice, the difference between a power law with a high exponent and an exponential function is so small that the results in terms of crack growth are similar. Equation (4.4) has the main advantage of being consistent with the kinematics of the chemical reaction that governs (according to the theory of *Charles* and *Hilling*) crack growth (Charles and Hilling, 1962; Wiederhorn et al., 1980). For practical use however, Equation (4.2) has the advantage of allowing for much simpler calculations. This explains its predominant use.

4.1.4 Chemical background

The classical stress corrosion theory of *Charles* and *Hilling* involves the chemical reaction of a water molecule with silica, which occurs at the crack tip (Charles and Hilling, 1962):

$$\text{Si-O-Si} + \text{H}_2\text{O} \rightarrow \text{Si-OH} + \text{HO-Si} \tag{4.5}$$

According to this theory, the crack velocity scales with the kinetics of this chemical reaction. Its activation energy depends on the local stress and on the radius of curvature at the crack tip. The theory involves a first order chemical process, which is consistent with the observed linear correlation between the logarithm of the crack velocity v and the logarithm of the humidity ratio H (except for very low H or v) (Wiederhorn and Bolz, 1970). *Wiederhorn* also showed that in region II, the kinetics of the chemical reaction at the crack tip are not controlled by the activation of the chemical process any more, but by the supply rate of water. It takes time for a water molecule to be transported to the crack tip. This causes a shortage in the supply of water as the crack velocity increases. A detailed chemical model, summarized in Figure 4.2, has been developed in Michalske and Freiman (1983).

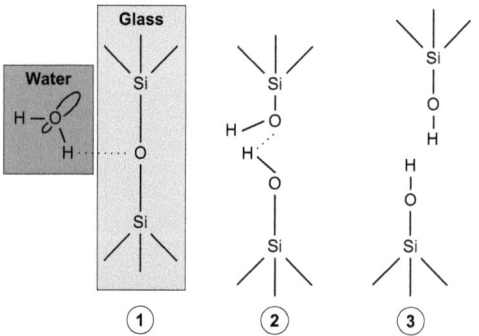

Figure 4.2:
Stress corrosion, chemical reaction at the crack tip: (1) adsorption of water to Si–O bond, (2) concerted reaction involving simultaneous proton and electron transfer, and (3) formation of surface hydroxyl groups (Michalske and Freiman, 1983).

This classical interpretation involving a chemical reaction at the very tip of a crack is questioned by *Tomozawa*. As the diffusion of molecular water into the glass is activated by stress, he suggests that this diffusion process and the modification of the glass properties in the crack tip area that it causes might explain subcritical crack growth (Tomozawa, 1998).

A more in-depth discussion is beyond the scope of this document. The interested reader should refer to the numerous texts available on this subject, e. g. Bunker (1994); Gehrke et al. (1990); Gy (2003); Michalske and Freiman (1983); Tomozawa (1998); Wiederhorn and Bolz (1970).

4.1.5 Crack healing, crack growth threshold and hysteresis effect

In the early days of research on the mechanical behaviour of glass it was found that aging has an effect on glass surface flaws (Levengood, 1958). Further experimental work was done in laboratory conditions by Wiederhorn and Tornsend (1970) and Stavrinidis and Holloway (1983). All experiments show that the strength of flawed specimens increases during stress-free phases. Looking at it in more detail, this effect, generally called *crack healing*, is a consequence of two phenomena, the crack growth threshold and the hysteresis effect.

At stress intensities below the *crack growth threshold* K_{th}, also known as 'stress corrosion limit', 'crack growth limit', 'threshold stress intensity', or 'fatigue limit', there is no more crack growth or it is so slow that it cannot be measured. The threshold appears strongly depend on the environmental conditions (e. g. on the pH value of a liquid the glass is immersed in) and on the glass's chemical composition. It is more easily evidenced with alkali containing glasses and in neutral or acidic environments. For typical soda lime silica glass at a moderate pH value, K_{th} is about 0.2 to 0.3 MPa m$^{0.5}$ (cf. Section 6.2). In alkali containing glasses, there is also a *hysteresis effect*: an aged crack will not repropagate immediately on reloading. The hysteresis effect is convincingly explained by renucleation of the aged crack in a plane different from the original one, as if the path of the crack has to turn around the area just in front of the former crack tip. This non-coplanar re-propagation was directly observed by atomic force microscopy (AFM) (Hénaux and Creuzet, 1997; Wiederhorn et al., 2002, 2003).

The crack growth threshold was originally explained by a rounding of the crack tip ('*crack tip blunting*') at slow crack velocities (Charles and Hilling, 1962). Direct evidence of crack tip blunting in pure silica glass was given by transmission electron microscopy (TEM) observation of the tip of a crack in a very thin film of silica glass aged in water (Bando et al., 1984). More recent investigations, however, strongly support the hypothesis that alkali are leached out of the glass and that this change in the chemical composition at the tip of the crack is responsible for the crack growth threshold rather than a geometrical change (blunting). AFM observations of aged indentation cracks did not give any evidence of blunting. Sodium containing crystallites were actually found on the surface of glass close to the tip of the indentation crack. This is more consistent with alkali ions' migration under the high stress at the crack tip and their exchange with protons or hydronium ions[3] from the environment (Gehrke et al., 1991; Guin and Wiederhorn, 2003; Nghiem, 1998).

A probabilistic crack propagation model, which takes these phenomena into account, is proposed in Charles et al. (2003).

Should healing or threshold effects be taken into account for structural glass applications?

Although its favourable influence can be considerable, crack healing has not been taken into account by design proposals to this day. The main problem is that crack healing is difficult to quantify and strongly dependent on the environmental conditions. While the crack growth threshold for SLS glass increases in acidic environments, there is no evidence of a threshold in alkaline environments (Green et al., 1999). In static long-term outdoor tests, in contrast to tests in the climatic chamber, no evidence of any substantial crack healing or of a crack growth threshold was found (Fink, 2000). *For structural applications, in which safety is a major concern, it therefore remains advisable not to take any threshold or healing effects into account.*

4.1.6 Influences on the relationship between stress intensity and crack growth

The following environmental conditions and material properties have a significant influence on the relationship between stress intensity and crack growth:

- **Humidity.** As shown in Figure 4.3a, the effect of an increasing water content of the surrounding medium is essentially a parallel shift of region I of the v-K_I relationship towards lower stress intensities and of region II towards higher crack velocities. This means that mainly the linear crack velocity parameter v_0 is affected while the slope of the curve, represented by the exponential crack velocity parameter n, does not change significantly. Newer results confirm this trend, see Section 6.3. It can be shown that it is not the absolute quantity of water that governs subcritical crack growth behaviour, but the ratio of the actual partial pressure to the partial pressure at saturation (Wiederhorn, 1967). For structural applications, this basically means that it is the relative humidity of the air that governs subcritical crack growth.[4]

- **Temperature.** Figure 4.3b shows schematic v-K_I-curves which were determined experimentally by Wiederhorn and Bolz (1970) for various temperatures. It can be seen that the main effect of an increasing temperature is again a parallel shift of the curve towards smaller K_I. Additionally, there is a slight decrease in the slope (n).

- **Corrosive media and pH value.** The crack velocity generally increases with an increasing pH value of the surrounding medium (Figure 4.3c). Furthermore, the pH value has a certain effect

---

[3] A hydronium ion is the cation H_3O^+ derived from protonation of water.
[4] In domains other than structural engineering, this issue is of more interest. A liquid that can only dissolve very small quantities of water may be close to saturation even at a low water content and can therefore not be considered to be inert even if the water content is very low.

on n and a particularly strong influence on the crack growth threshold K_{th}. More detailed information can be found e. g. in Gehrke et al. (1990).

- **Chemical composition of the glass.** All parameters of subcritical crack growth are influenced by the chemical composition of the glass (Figure 4.3d).

Because of crack healing (see Section 4.1.5), the **age of the flaws** has an influence on the onset of crack growth.

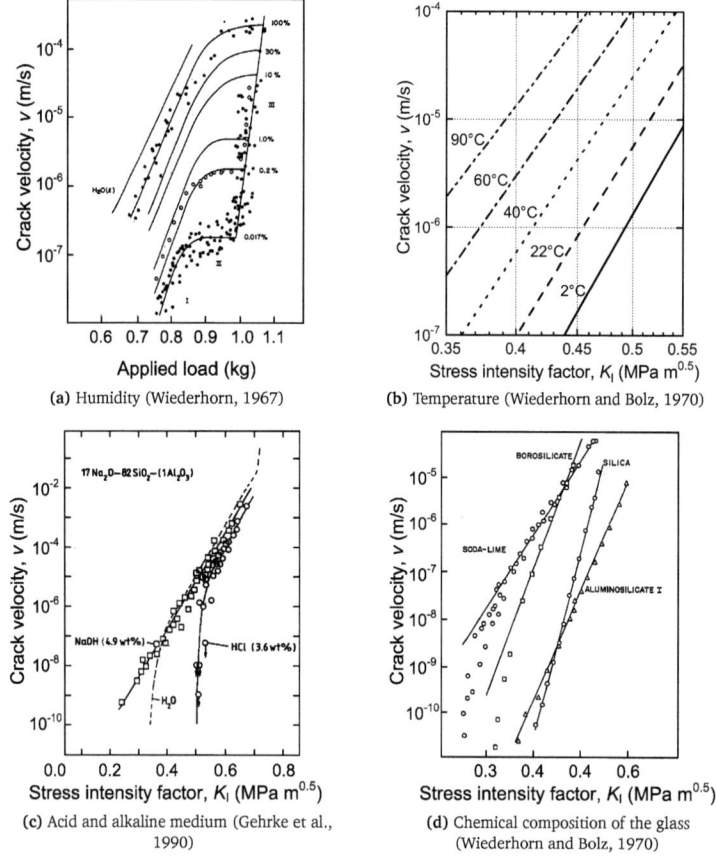

Figure 4.3: Influences on subcritical crack growth.

4.2 Mechanical behaviour of a single surface flaw

4.2.1 Fracture mechanics basics

Glass is an almost perfectly elastic material. Therefore linear elastic fracture mechanics (LEFM) is an ideal theory to model its behaviour. In fact, glass was the material used for the development of the basis of LEFM. In LEFM, mechanical material behaviour is modelled by looking at *cracks*. A crack is an idealized model of a flaw having a defined geometry and lying in a plane. It may either be located on the surface (*surface crack*) or embedded within the material (*volume crack*). For structural glass elements, only surface cracks need to be considered (Figure 4.4).

In 1920, *Griffith* postulated a fracture concept developed from experiments on glass specimens and based on the consideration of the surface energy in the region surrounding the crack (Griffith, 1920). *Irwin* (Irwin, 1957) and others developed the concept of the stress intensity factor (SIF) based on this fundamental work. This concepts allows the basic rule of glass failure to be expressed in simple terms: *A glass element fails, if the stress intensity factor K_I due to tensile stress at the tip of one crack reaches its critical value K_{Ic}.*

The general relationship between the *stress intensity factor* K_I, the nominal tensile stress normal to the crack's plane σ_n, a correction factor Y, and some representative geometric parameter a, in general the crack depth or half of the crack length, is given by:[5]

$$K_I = Y \cdot \sigma_n \cdot \sqrt{\pi \cdot a} \qquad (4.6)$$

The *fracture toughness*[6] K_{Ic}, also known as the critical stress intensity factor, is the SIF that leads to instantaneous failure. This is known as *Irwin's fracture criterion*:

$$K_I \geq K_{Ic} \qquad (4.7)$$

The criterion assumes pure mode I' (opening mode) fracture of a crack exposed to uniaxial tension normal to the crack's plane. More general cases are discussed in Section 4.3.3.

The correction factor Y[7] in Equation (4.6) depends on the stress field, the crack depth, the crack geometry and the element geometry. The dependence on the element geometry, the stress field and the crack depth is small for shallow surface cracks and can generally be neglected (see Section 6.2). The dependence on the crack geometry is more significant, which is why Y is called the *geometry factor* henceforth.

Figure 4.4:
Fundamental terms used to describe surface cracks.

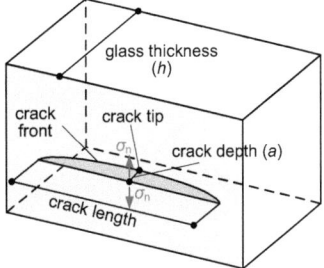

---

[5] Stress intensity considerations for simple cases are usually based on the assumption of a homogeneous distribution of the nominal stress within the section. As long as the stress variation in the proximity of the crack is small, which is the case for cracks with a small depth in comparison to the material thickness, this assumption is a good approximation even for non-homogeneous' stress fields.

[6] It should be noted that the 'c' in K_{Ic} refers to the word 'critical' and not to the crack length c as it is often the case in the context of fatigue of steel structures.

[7] Caution when using published data: In many publications Y is used as a synonym for $Y\sqrt{\pi}$.

Excursus: Energy release rate

In scientific publications, the energy release rate G is often used instead of the stress intensity factor K_I. For an elastic material and crack mode I, it is

$$G = \frac{K_I^2}{E'} \qquad E' = \begin{cases} E & \text{for plane stress state} \\ E/(1-v^2) & \text{for plane strain state} \end{cases} \qquad (4.8)$$

where E is Young's modulus and v is Poisson's ratio (Irwin, 1957). In the case of shallow surface cracks, a plane stress state may be assumed.

4.2.2 Inert strength

With Equation (4.6) and Equation (4.7), a crack extends immediately to cause failure if:

$$Y \cdot \sigma_n \cdot \sqrt{\pi \cdot a} \geq K_{Ic} \qquad (4.9)$$

With this, the stress causing failure of a crack of depth a, the *critical stress* σ_c, is

$$\sigma_c(t) = \frac{K_{Ic}}{Y \cdot \sqrt{\pi \cdot a(t)}} \qquad (4.10)$$

while the depth of a crack failing at some stress σ_n, the *critical crack depth* a_c, is

$$a_c(t) = \left(\frac{K_{Ic}}{\sigma_n(t) \cdot Y \sqrt{\pi}} \right)^2. \qquad (4.11)$$

Both the stress σ_n and the crack depth a are time-dependent. Therefore σ_c and a_c are also time-dependent. The critical stress represents the strength of a crack when no subcritical crack growth occurs and is therefore called *inert strength* henceforth. It is plotted in Figure 4.5 as a function of the crack depth using typical parameters for a long, macroscopic surface crack of small depth in a glass plate.

For information on how to measure the inert strength in laboratory tests, see Section 7.3.

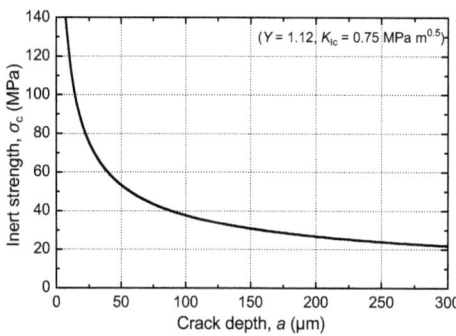

Figure 4.5:
Inert strength of a single crack as a function of its depth (for information on the values of Y and K_{Ic}, see Chapter 6).

4.2. MECHANICAL BEHAVIOUR OF A SINGLE SURFACE FLAW

4.2.3 Heat treated glass

The term *heat treated glass* includes *any* glass type that has been processed in order to induce residual stresses (cf. Section 2.3.2), namely heat strengthened glass and fully tempered glass.

The in-plane surface stress normal to a crack's plane (index 'n'), also known as the *crack opening stress*, is:

$$\sigma_n(\tau,\vec{r},\varphi) = \sigma_{E,n}(\tau,\vec{r},\varphi) + \sigma_{r,n}(\vec{r},\varphi) + \sigma_{p,n}(\tau,\vec{r},\varphi) \tag{4.12}$$

σ_E surface stress due to actions
σ_r residual surface stress due to tempering ('prestress')
σ_p surface stress due to external constraints or prestressing
τ point in time
$\vec{r}$ crack location
φ crack orientation

A crack can only grow or fail if it is exposed to tensile stress, i.e. if $\sigma_n(t,\vec{r},\varphi) > 0$. By considering negative σ_n as $\sigma_n = 0$ in crack growth calculations, the effect of residual stresses can be accounted for in a simple and consistent way and the same design equations or algorithms can be used for all glass types. If a heat treated glass element is designed such that (T is the design life)

$$\max_{\tau \in [0,T]} \sigma_{E,n} \leq -(\sigma_{r,n} + \sigma_{p,n}) \; \forall (\vec{r},\varphi), \tag{4.13}$$

no surface decompression occurs, i.e. no surface crack is ever exposed to tensile stress during the entire service life. Such an element does not show any size-dependent, time-dependent or environment-dependent effects.

4.2.4 Subcritical crack growth and lifetime

Assuming the ordinary differential equation of crack growth (cf. Section 4.1.3)

$$v = \frac{da}{dt} = v_0 \left(\frac{K_I}{K_{Ic}}\right)^n \tag{4.14}$$

to be valid over the full range of K_I (which means neglecting the crack growth threshold), using the stress intensity factor from Equation (4.6) and assuming n to be constant, variable separation yields

$$\int_{a_i}^{a(t)} a^{-\frac{n}{2}} \, da = \int_0^t v_0 \cdot K_{Ic}^{-n} \cdot \left(Y\sqrt{\pi}\right)^n \cdot \sigma_n^n(\tau) \, d\tau \tag{4.15}$$

with a_i being the initial crack depth ($a_i = a(t=0)$). The *time-dependent size of a single crack* exposed to the crack opening stress history $\sigma(t)$ is thus:

$$a(t) = \left[a_i^{\frac{2-n}{2}} + \frac{2-n}{2} \cdot v_0 \cdot K_{Ic}^{-n} \cdot \left(Y\sqrt{\pi}\right)^n \cdot \int_0^t \sigma_n^n(\tau) \, d\tau \right]^{\frac{2}{2-n}} \tag{4.16}$$

Variable separation and integration over the time interval $[0,T]$ and the corresponding crack depths $[a_i, a]$ gives the following basic relationship:

$$\int_0^T \sigma_n^n(\tau) \, d\tau = \frac{2}{(n-2) \cdot v_0 \cdot K_{Ic}^{-n} \cdot (Y\sqrt{\pi})^n \cdot a_i^{(n-2)/2}} \left[1 - \left(\frac{a_i}{a}\right)^{(n-2)/2}\right] \tag{4.17}$$

The crack depth at failure a_f is the critical crack depth (Equation (4.11)) for the failure stress $\sigma(t_\mathrm{f})$

$$a_\mathrm{f} = \left(\frac{K_\mathrm{Ic}}{\sigma_\mathrm{n}(t_\mathrm{f}) \cdot Y\sqrt{\pi}}\right)^2 \tag{4.18}$$

with t_f being the time to failure or *lifetime* of the crack in question. This can now be inserted into Equation (4.17). As n is large (≈ 16), the expression in square brackets in Equation (4.17) approaches 1 for long lifetimes with $a_\mathrm{f} \gg a_\mathrm{i}$. Thus the following simplified expression may be obtained:

$$\int_0^{t_\mathrm{f}} \sigma_\mathrm{n}^n(\tau)\,\mathrm{d}\tau = \frac{2}{(n-2) \cdot v_0 \cdot K_\mathrm{Ic}^{-n} \cdot (Y\sqrt{\pi})^n \cdot a_\mathrm{i}^{(n-2)/2}} \tag{4.19}$$

This is the classic relationship of current glass design. Given a stress history, it enables the calculation of the lifetime of a crack given its initial depth or the allowable initial crack depth given its required lifetime. The left hand side of Equation (4.19) is called *risk integral* or *'Brown's integral'*, because it was first used by *Brown* (Brown, 1972) to characterize damage accumulation in glass.

While Equation (4.19) is very convenient, it suffers from its limit of validity. If crack velocity is slow and/or the loading time is very short (near-inert conditions), the crack depth at failure may not be much bigger than the initial crack depth. In fact, in perfectly inert conditions, both depths are identical. The assumption of $a_\mathrm{f} \gg a_\mathrm{i}$ used to obtain Equation (4.19) is therefore *not* valid in such conditions. From Equation (4.16) and Equation (4.18), a formulation of general validity can be obtained:

$$\tilde{a}_c(\tau) = \left[\left(\frac{\sigma_\mathrm{n}(\tau) \cdot Y\sqrt{\pi}}{K_\mathrm{Ic}}\right)^{n-2} + \frac{n-2}{2} \cdot v_0 \cdot K_\mathrm{Ic}^{-n} \cdot (Y\sqrt{\pi})^n \cdot \int_0^\tau \sigma_\mathrm{n}^n(\tilde{\tau})\,\mathrm{d}\tilde{\tau}\right]^{\frac{2}{2-n}} \tag{4.20}$$

The crack depth $\tilde{a}_c(\tau)$ is the initial depth of a crack that fails at the point in time τ when exposed to the crack-opening stress history $\sigma_\mathrm{n}(\tau)$. The choice of the symbol will become clear in Section 4.3. The disadvantage of Equation (4.20) is that it depends not only on the risk integral but also on the momentary stress $\sigma_\mathrm{n}(\tau)$. While the risk integral is monotonously increasing', in general the momentary stress is not. Therefore, the minimum initial crack depth $\min(\tilde{a}_c(\tau))$, which is relevant for design, does not necessarily occur at the end of the stress history ($\tau = T$) but may occur at any $\tau \in [0, T]$. A crack does not fail if

$$a_\mathrm{i} < \min_{\tau \in [0,T]} \tilde{a}_c(\tau). \tag{4.21}$$

The behaviour of a single surface flaw is discussed quantitatively in Section 8.3.1. The problem of quantities that do not only depend on the risk integral but on the entire stress history is further discussed in Section 5.2.1.

4.2.5 Equivalent static stress and resistance

It will be seen in Section 5.2.1 that the simplified expression in Equation (4.19) is generally sufficient for structural design (but not for the interpretation of experiments). This equation means that, if n is constant, two stress histories $\sigma_{(1)}(\tau)\ \tau \in [0, t_1]$ and $\sigma_{(2)}(\tau)\ \tau \in [0, t_2]$ cause the same crack growth if $\int_0^{t_1} \sigma_{(1)}^n(\tau)\,\mathrm{d}\tau = \int_0^{t_2} \sigma_{(2)}^n(\tau)\,\mathrm{d}\tau$. The value of these integrals increases from 0 at the beginning of the loading to the value of Equation (4.19) at failure. It is convenient for subsequent chapters to define an *equivalent stress* σ_T, which is the stress that would cause the same crack growth when constantly applied during T, as the stress history $\sigma(\tau)$ with $\tau \in [0, T]$: $\int_0^T \sigma^n(\tau)\,\mathrm{d}\tau = \sigma_T^n \cdot T$. Any stress history can thus be characterized by a single equivalent stress value σ_{t_0} that would, when acting during the *reference time period* t_0, have the same effect as the original stress history:

4.2. MECHANICAL BEHAVIOUR OF A SINGLE SURFACE FLAW

$$\sigma_{t_0} = \left(\frac{1}{t_0} \int_0^T \sigma^n(\tau) \, d\tau \right)^{1/n} \approx \left(\frac{1}{t_0} \sum_{j=1}^J \left[\sigma_{t_j}^n \cdot t_j \right] \right)^{1/n} \quad (4.22)$$

The stress σ_{t_0} is called t_0-*equivalent stress*. The right side of Equation (4.22) caters for discrete stress histories consisting of J time periods of duration t_j ($T = \sum t_j$) and constant stress σ_{t_j}.

The same approach can be used for a crack's resistance by defining the t_0-*equivalent resistance*:

$$\sigma_{R,t_0} = \left(\frac{1}{t_0} \frac{2}{(n-2) \cdot v_0 \cdot K_{Ic}^{-n} \cdot (Y\sqrt{\pi})^n \cdot a_i^{(n-2)/2}} \right)^{1/n} \quad (4.23)$$

This is the static stress that a crack can resist for a reference time period t_0 (usually $t_0 = 1\,\mathrm{s}$, $3\,\mathrm{s}$ or $60\,\mathrm{s}$). It is independent of the load and completely characterizes the load capacity of a given crack (or an element whose load capacity is governed by this crack) for given environmental conditions (v_0, n), initial crack depth (a_i) and crack geometry (Y). A structural safety verification based on this approach entails ensuring that:[8]

$$\sigma_{t_0} \leq \sigma_{R,t_0} \quad (4.24)$$

Excursus 1: Calculation of the equivalent stress for common stress histories

Menčík proposed a useful variation of Equation (4.22), expressing the equivalent stress σ_T (see above) in terms of a chosen, characteristic value σ_{ch} and a shape coefficient g (Menčík, 1992):

$$\sigma_T = g^{1/n} \sigma_{ch} \quad \text{with} \quad g = T^{-1} \int_0^T \left(\frac{\sigma(\tau)}{\sigma_{ch}} \right)^n d\tau \quad (4.25)$$

For a constant stress σ_{const}, the coefficients are $\sigma_{ch} = \sigma_{const}$ and $g = 1$. For a constant stress rate ($\sigma(\tau) = \dot{\sigma}_{const} \cdot \tau$), it is $g = 1/(n+1)$ and $\sigma_{ch} = \max(\sigma(\tau))$.

Excursus 2: Using equivalent time periods instead of equivalent stresses

Menčík furthermore proposed to define equivalent quantities by standardizing to the unit stress $\sigma_{01} = 1\,\mathrm{MPa}$ instead of a reference time period (Menčík, 1984). This allows for the definition of the *normalized loading time* t_n as

$$t_n = \int_0^T \left(\frac{\sigma(\tau)}{\sigma_{01}} \right)^n d\tau = \sum_{j=1}^J \left(\frac{\sigma_{t_j,j}}{\sigma_{01}} \right)^n \cdot t_j = \left(\frac{\sigma_{t_0}}{\sigma_{01}} \right)^n t_0 \quad (4.26)$$

and of the *normalized lifetime* $t_{f,n}$ (index f from 'failure') as

$$t_{f,n} = \int_0^{t_f} \left(\frac{\sigma(\tau)}{\sigma_{01}} \right)^n d\tau = \frac{1}{\sigma_{01}^n} \frac{2}{(n-2) \cdot v_0 \cdot K_{Ic}^{-n} \cdot (Y\sqrt{\pi})^n \cdot a_i^{(n-2)/2}} \cdot \quad (4.27)$$

The physical meaning of this relationship is that the unit stress σ_{01} acting for t_n causes the same crack growth as the actual stress history $\sigma(\tau)$ with $\tau \in [0, T]$. A crack of initial depth a_i that is exposed to σ_{01} fails after $t_{f,n}$. This approach can easily be related to the equivalent stress concept:

$$t_n = t_0 \left(\frac{\sigma_{t_0}}{\sigma_{01}} \right)^n \quad ; \quad t_{f,n} = t_0 \left(\frac{\sigma_{R,t_0}}{\sigma_{01}} \right)^n \quad (4.28)$$

---

[8] In ASTM E 1300-04, this approach is used for entire glass plates instead of for a single crack and for lateral load instead of for surface stresses. This yields the so-called 'non-factored load', which is actually the 3 s-equivalent resistance of a glass plate of defined geometry and support conditions to uniform lateral load. Details are discussed in Chapter 3, drawbacks and limits of applicability in Section 5.3.2.

Excursus 3: Damage accumulation and residual lifetime

Like all time quantities, normalized times can be added and subtracted. Therefore, the *normalized residual lifetime* $t_{\text{res,n}}$ at a point in time t is

$$t_{\text{res,n}}(t) = t_{\text{f,n}} - t_{\text{n}}(t). \tag{4.29}$$

Division by $t_{\text{f,n}}$ yields the dimensionless *relative residual lifetime* D_{res}:

$$D_{\text{res}}(t) = 1 - \frac{t_{\text{n}}(t)}{t_{\text{f,n}}} \tag{4.30}$$

This quantity is equal to 1 at the beginning of the loading and progressively decreases towards 0 until failure. It is thus useful as a limit state function or safety margin. The term $t_{\text{n}}(\tau)/t_{\text{f,n}}$ is, analogous to the relative damage commonly used for fatigue calculations of steel structures, the *relative reduction of lifetime*. Finally, the relative residual lifetime can be expressed in terms of equivalent stresses or of the momentary crack depth $a(t)$:[9]

$$D_{\text{res}}(t) = 1 - \left(\frac{\sigma_{t_0}}{\sigma_{R,t_0}}\right)^n = \left(\frac{a_i}{a(t)}\right)^{\frac{n-2}{2}} \tag{4.31}$$

D_{res} does not depend on the actual values of the normalization variables σ_{01} and t_0.

4.2.6 Constant stress and constant stress rate

The relationship between lifetimes and applied constant stresses of two identical cracks (a_i, Y) in identical conditions (v_0, n, K_{Ic}) follows directly from Equation (4.22):

$$\frac{\sigma_2}{\sigma_1} = \left(\frac{t_1}{t_2}\right)^{1/n} \quad \text{or} \quad \frac{t_1}{t_2} = \left(\frac{\sigma_2}{\sigma_1}\right)^n \tag{4.32}$$

Inserting $\sigma(\tau) = \dot{\sigma}_{\text{const}} \cdot \tau$ into Equation (4.22), the relationship between the lifetime of two identical cracks (a_i, Y) in identical conditions (v_0, n, K_{Ic}) loaded at constant stress rates $\dot{\sigma}_1$ and $\dot{\sigma}_2$ is obtained:

$$\frac{t_1}{t_2} = \left(\frac{\dot{\sigma}_2}{\dot{\sigma}_1}\right)^{\frac{n}{n+1}} \tag{4.33}$$

The fact that these equations are independent of v_0 has certain advantages. The failure stress as a function of the stress rate has a slope of $1/(n+1)$ when plotted on logarithmic scales. This allows the parameter n to be determined from experiments with variable stress rate. It should not be overlooked, however, that while the equations are independent of v_0, their validity is confined to cases in which flaws and conditions, *including* v_0, are identical during all tests. Since v_0 can be strongly stress rate dependent (see Chapter 6), this method should be used with caution.

---

[9] Using Equations (4.19), (4.28), (4.30) and (4.31).

4.3 Extension to a random surface flaw population

4.3.1 Starting point and hypotheses

The model in Section 4.2, (Equations (4.19) and (4.20)), is based on the growth of a single crack. For several cases of practical relevance, this is an appropriate way of modelling the surface condition of structural glass elements. For others, however, it is not. The issue will be discussed in detail in Chapter 8. For the moment it is sufficient to consider a simple, yet common, case in which a single flaw model is inappropriate, namely as-received glass. When glass is delivered to the client by manufacturers, it does normally not contain any flaws that are clearly deeper than others and would thus be certain to fail first upon loading. As stated in Section 2.2.2, the surface contains a large number of mechanical flaws of varying severity, which are not necessarily visible to the naked eye. This surface condition can much better be represented by a *random surface flaw population* (RSFP). It is therefore necessary to extend the single flaw model to describe glass elements in which resistance is governed by an RSFP.

The derivation is based on several hypotheses:

1. The material contains a large number of natural flaws of variable depth.
2. The depth of the flaws is a random variable and can be described using a probability distribution function.[10]
3. The individual flaws do not influence each other.[11]
4. The element fails when the first flaw fails (weakest-link model).

In order to make the derivation as clear and understandable as possible, additional simplifying assumptions are made, namely:

1. The orientation of all flaws is identical and perpendicular to the homogeneous tensile stress σ.
2. There is no subcritical crack growth.

Taking these very restrictive assumptions as a starting point, the model is generalized step by step, omitting one simplifying restriction after the other to finally obtain the most general solution possible. Inevitably, some hypotheses remain integral parts of the model. Chapter 5 discusses the validity of these hypotheses as well as possible simplifications of the lifetime prediction model and the conditions that they are valid for.

In addition to linear elastic fracture mechanics, about which some key publications have already been cited in Section 4.2, the present section makes use of fundamental work in the fields of theory of probability and strength of materials including Weibull (1939, 1951), Batdorf and Crose (1974), Batdorf and Heinisch (1978) and Evans (1978).

4.3.2 Constant, uniform, uniaxial stress

With a constant stress and no subcritical crack growth, the crack depth a and the critical crack depth a_c (cf. Section 4.2.2) are both constant. The failure probability of a crack is simply the probability that its random size a is larger than the critical crack depth a_c:

$$P_{f,\text{inert}}^{(1)}(a) = \mathscr{P}\left(a \geq a_c\right) = \int_{a_c}^{\infty} f_a(a)\,\mathrm{d}a = 1 - F_a\left(a_c\right) \tag{4.34}$$

---

[10] One could also consider the *strength* of the flaws as the basic random variable. The choice is irrelevant because both quantities can be expressed in terms of each other using linear elastic fracture mechanics. The relationship between the two cumulative distribution functions is found from Equation (4.10) and Equation (4.11): $F_a(a_c) = \mathscr{P}(a < a_c) = \mathscr{P}\left(\left[K_{\text{Ic}}/(\sigma_c \cdot Y\sqrt{\pi})\right]^2 < \left[K_{\text{Ic}}/(\sigma \cdot Y\sqrt{\pi})\right]^2\right) = \mathscr{P}(\sigma_c > \sigma) = 1 - \mathscr{P}(\sigma_c < \sigma) = 1 - F_{\sigma_c}(\sigma) = 1 - F_{\sigma_c}\left[K_{\text{Ic}}/(Y\sqrt{\pi}\sqrt{a_c})\right]$

[11] This assumption is conservative. The presence of cracks modifies the stress field within the material. If the length of a surface crack is similar to, or longer than, the distance separating cracks, it induces a shielding of the stress at the neighbouring crack tip. This effect can reduce crack growth and increase lifetime (Auradou et al., 2001).

F_a is the cumulative distribution function (CDF), f_a the probability density function (PDF) of the crack depth. The strength distribution mainly depends on the distribution of the larger flaws. Assuming that the mean number of flaws is large (cf. Section 4.3.1), the mathematical theory of extreme values applies and shows that the asymptotic behaviour of the crack depth distribution can be described accurately by a power law. The probability density function (PDF) of the crack depth a is thus ($\propto$ means 'proportional to', r is a parameter):

$$f_a(a) \propto \frac{1}{a^r} \qquad (4.35)$$

The CDF of the crack depth $F_a = \int f_a$ is a *Pareto distribution*[12]:

$$F_a(a) = \begin{cases} 0 & \text{for } a \leq a_0 \\ 1 - (a_0/a)^{r-1} & \text{for } a > a_0 \end{cases} \qquad (4.36)$$

For normalization reasons of the CDF ($F_a = 1$ for $a \to \infty$), a lower limit a_0 for the crack depth a has to be introduced. Since very small cracks are irrelevant for failure, the actual value of a_0 is unimportant. Equation (4.36) sufficiently describes the crack distribution in the range of relevant crack depths.

An element fails if *any* of the flaws fail, or survives if *all* flaws survive. The *survival probability* of one flaw being

$$P^{(1)}_{s,\text{inert}} = 1 - P^{(1)}_{f,\text{inert}}, \qquad (4.37)$$

the survival probability of an element with exactly k flaws is the product of the survival probabilities of all flaws:

$$P^{(k)}_{s,\text{inert}} = \left(1 - P^{(1)}_{f,\text{inert}}\right)^k \qquad (4.38)$$

In a real glass element, the total number of flaws is a random variable itself and k remains unknown. Therefore, only the *mean* number of flaws M shall be considered as a known variable. If the real number of flaws follows a Poisson distribution with mean M, the probability $P^{(k)}$ of an element having exactly k flaws is given by:

$$P^{(k)} = \frac{e^{-M} \cdot M^k}{k!} \qquad (4.39)$$

The survival probability of a random glass element is now obtained by multiplying the probability $P^{(k)}$ of an element of surface area A having exactly k flaws by the corresponding survival probability from Equation (4.38) and summing up over all possible numbers of flaws:

$$P_s = \sum_{k=0}^{\infty} P^{(k)} \cdot P^{(k)}_{s,\text{inert}} \qquad (4.40)$$

Using the definition of the exponential function as an infinite series

$$e^x = \sum_{k=0}^{\infty} \frac{1}{k!} x^k, \qquad (4.41)$$

the survival probability becomes

$$P_{s,\text{inert}} = e^{\left(-M \cdot P^{(1)}_{f,\text{inert}}\right)} \qquad (4.42)$$

which finally gives the failure probability as a function of the critical crack depth a_c:

$$P_{f,\text{inert}}(a_c) = 1 - P_{s,\text{inert}} = 1 - \exp\{-M\left[1 - F_a(a_c)\right]\} = 1 - \exp\left\{-M\left(\frac{a_0}{a_c}\right)^{r-1}\right\} \qquad (4.43)$$

---

[12] For a visualization of the Pareto distribution, see Figure 5.4.

4.3. EXTENSION TO A RANDOM SURFACE FLAW POPULATION

Inserting a_c from Equation (4.11), this can be expressed in terms of the momentary stress σ:

$$P_{f,\text{inert}}(\sigma) = 1 - \exp\left\{-M a_0^{r-1} \cdot \left(\frac{\sigma Y \sqrt{\pi}}{K_{\text{Ic}}}\right)^{2(r-1)}\right\}$$

$$= 1 - \exp\left\{-\left(\frac{\sigma}{\vartheta}\right)^{m_0}\right\} \tag{4.44}$$

$$\vartheta = \frac{K_{\text{Ic}}}{M^{1/m_0} \cdot Y \sqrt{\pi} \cdot \sqrt{a_0}} \qquad m_0 = 2(r-1) \tag{4.45}$$

As ϑ contains the total number of flaws M, it depends on an element's surface area A. In order to avoid this, M can be expressed in terms of the mean number of flaws M_0 on the unit surface area A_0:

$$M = \frac{M_0}{A_0} \cdot A \tag{4.46}$$

This finally yields the *inert failure probability* $P_{f,\text{inert}}$ of a glass element:

$$P_{f,\text{inert}}(\sigma) = 1 - \exp\left\{-\frac{A}{A_0}\left(\frac{\sigma}{\theta_0}\right)^{m_0}\right\} \tag{4.47}$$

$$\theta_0 = \frac{K_{\text{Ic}}}{M_0^{1/m_0} \cdot Y \sqrt{\pi} \cdot \sqrt{a_0}} \qquad m_0 = 2(r-1) \tag{4.48}$$

The parameters θ_0 and m_0 *solely* depend on the surface flaw population and are therefore *true material parameters*. They can be measured in inert tests. High values of m_0 represent a narrow distribution of the crack depths and therefore of inert strengths. High values of θ_0 represent a high mean and wide distribution of inert strengths. Equation (4.47) can be rearranged to take on the form of a *two-parameter Weibull distribution* with scale parameter θ_{inert} and shape parameter m_0:

$$P_{f,\text{inert}}(\sigma) = 1 - \exp\left\{-\left(\frac{\sigma}{\theta_{\text{inert}}}\right)^{m_0}\right\} \tag{4.49}$$

$$\theta_{\text{inert}} = \theta_0 \left(\frac{A}{A_0}\right)^{-1/m_0} \tag{4.50}$$

For two elements with surface areas A_1 and A_2 exposed to tensile stress, it is:

$$\frac{\theta_{\text{inert},A_1}}{\theta_{\text{inert},A_2}} = \left(\frac{A_2}{A_1}\right)^{1/m_0} \tag{4.51}$$

This ratio is commonly referred to as *size effect*. When replacing $\sigma(A_1)/\sigma(A_2)$ by $\theta_{\text{inert},A_1}/\theta_{\text{inert},A_2}$ and s by m_0, Figure 3.9 visualizes Equation (4.51). It can be seen that the size effect is small for high values of m_0 (small scatter of strength) and large for small values of m_0 (large scatter).

Excursus: Reference inert strength

From Equation (4.49), a quantity can be defined that will be useful when discussing design issues. This quantity $f_{0,\text{inert}}(P_{f,t}, \theta_0, m_0)$, called *reference inert strength* hereafter, is defined as follows: A glass surface of area $A_0 = 1\,\text{m}^2$ at inert conditions fails with probability $P_{f,t}$ when exposed to a uniformly distributed crack opening surface stress $f_{0,\text{inert}}$.

$$f_{0,\text{inert}}(P_{f,t}, \theta_0, m_0) = \theta_0 \left[-\ln(1 - P_{f,t})\right]^{1/m_0} \tag{4.52}$$

The reference inert strength depends on the target failure probability and the glass surface condition *only*. It does *not* depend on the glass type (because it refers to the crack opening stress) or on crack velocity parameters (because it refers to inert conditions).

4.3.3 Extension to non-uniform, biaxial stress fields

In a *non-uniform stress field*, the stress σ depends on the point on the surface $\vec{r} = (x, y)$. Equation (4.47) can be extended accordingly by integrating over infinitesimal surface elements dA of constant stress:

$$P_{\text{f,inert}}(\sigma) = 1 - \exp\left\{ -\frac{1}{A_0} \int_A \left(\frac{\sigma(\vec{r})}{\theta_0} \right)^{m_0} dA \right\} \quad (4.53)$$

Thus far, all cracks have been assumed to lie in a plane normal to the principal stress. Now, the model is to be extended to account for random crack orientation in a *biaxial stress field*. Implications of this, as well as possibilities for simplification and their limits of applicability, are discussed in Section 5.2.3.

Choosing a multimodal failure criterion

To assess the failure probability in a multiaxial stress field, the failure criterion needs to be generalized. For multimodal failure in a multiaxial stress field, a general failure criterion would be

$$g(K_{\text{I}}, K_{\text{II}}, K_{\text{III}}) \geq g_c. \quad (4.54)$$

K_{I}, K_{II} and K_{III} are the mode I, II and III stress intensity factors. The critical value g_c characterizes a material's resistance against unstable crack propagation and is a function of the fracture toughness K_{Ic} and possibly of K_{IIc} and K_{IIIc}. A mode I-equivalent stress intensity factor $K_{\text{I-eq}}$ is defined as the SIF that gives the same crack behaviour under pure mode I loading as the combination of all three SIFs give under multimodal loading:

$$g(K_{\text{I-eq}}, 0, 0) = g(K_{\text{I}}, K_{\text{II}}, K_{\text{III}}) \quad (4.55)$$

With the mode I geometry factor Y_{I}, the mode I-equivalent stress is:

$$\sigma_{\text{I-eq}} = \frac{K_{\text{I-eq}}}{Y_{\text{I}} \sqrt{\pi \cdot a}} \quad (4.56)$$

The critical crack depth from Equation (4.11) becomes

$$a_c = \left(\frac{K_{\text{Ic}}}{\sigma_{\text{I-eq}} \cdot Y_{\text{I}} \sqrt{\pi}} \right)^2 \quad (4.57)$$

and depends, via $\sigma_{\text{I-eq}}$, on the crack location and orientation.

Many failure criteria (g_c and $K_{\text{I-eq}}$) have been proposed and discussed in scientific publications[13]. By coaxial double ring and Brazilian disc tests, among others on alumina (Al$_2$O$_3$), which exhibit a fracture behaviour comparable to that of glass, it was shown that the simplest failure criterion, pure mode I fracture, gives clearly the best agreement with experimental results (Brückner-Foit et al., 1997). This criterion is therefore used and effects from mode II and mode III loading are ignored. Equation (4.54) and Equation (4.55) become

$$g(K_{\text{I}}, K_{\text{II}}, K_{\text{III}}) = K_{\text{I}} \quad \text{and} \quad g_c = K_{\text{Ic}}. \quad (4.58)$$

A crack therefore fails because of unstable crack propagation if $K_{\text{I}} > K_{\text{Ic}}$. The mode I-equivalent stress $\sigma_{\text{I-eq}}$ is equal to the stress component perpendicular to the crack σ_n:

$$\sigma_{\text{I-eq}} = \sigma_n \quad (4.59)$$

To simplify notation, the mode-I geometry factor will hereafter be termed Y (without the index I). With this notation, Equation (4.11) remains valid even for biaxial stress fields.

---

[13] For a summary see e. g. Thiemeier et al. (1991) or Brückner-Foit et al. (1997).

4.3. EXTENSION TO A RANDOM SURFACE FLAW POPULATION

Extending the failure probability equation

For a surface crack of orientation φ in plane stress state ($\sigma_z = \tau_{zx} = \tau_{zy} = 0$), the stress component normal to the crack is (Sayir, 1996)

$$\sigma_n = \sigma_x \cos^2\varphi + \sigma_y \sin^2\varphi + 2\tau_{xy} \sin\varphi \cos\varphi \qquad (4.60)$$

$$\tau = \left| \frac{\sigma_x - \sigma_y}{2} \cdot \sin(2\varphi) - \tau_{xy} \cdot \cos(2\varphi) \right| \qquad (4.61)$$

which, when expressed in terms of principal stresses, simplify to:

$$\sigma_n = \sigma_1 \cos^2\varphi + \sigma_2 \sin^2\varphi \qquad (4.62)$$

$$\tau = \frac{1}{2} |\sigma_1 - \sigma_2| \cdot \sin(2\varphi) \qquad (4.63)$$

σ_1, σ_2 major and minor in-plane principal stress ($\sigma_1 \geq \sigma_2$)

φ crack orientation ($\varphi = 0 \Rightarrow$ crack $\| \sigma_1$)

Analogous to the uniaxial case in Equation (4.34), but with a location-dependent and orientation-dependent critical crack depth $a_c(\vec{r}, \varphi)$, the failure probability of a single crack is:

$$P^{(1)}_{\text{f,inert}} = 1 - F_a\left(a_c(\vec{r}, \varphi)\right) \qquad (4.64)$$

Since glass is a homogeneous, isotropic material, it may be assumed that all crack locations $\vec{r} = (x, y)$ and crack orientations φ have the same probability of occurrence as long as no directional scratching is introduced (see Section 5.1.3). The probability density functions for a crack's location and orientation are thus both *uniform distributions*:

$$f_A(\vec{r}) = \frac{1}{A} \quad ; \quad f_\varphi(\varphi) = \frac{1}{\pi} \qquad (4.65)$$

The probability of finding a crack of orientation φ within the infinitesimally small surface area dA at the point $\vec{r}$ on the surface is

$$P_{\varphi,\vec{r}} = \frac{1}{A} dA \cdot \frac{1}{\pi} d\varphi. \qquad (4.66)$$

The probabilities $P^{(1)}_{\text{f,inert}}$ and $P_{\varphi,\vec{r}}$ need to be multiplied and all possible crack orientations[14] and locations must be summed to determine the probability $Q^{(1)}$ that a single crack leads to failure considering the dependence on location and orientation:

$$Q^{(1)} = \frac{1}{A} \int_A \frac{2}{\pi} \int_{\varphi=0}^{\pi/2} \left(1 - F_a(a_c(\vec{r}, \varphi))\right) dA \, d\varphi \qquad (4.67)$$

Using F_a from Equation (4.36), this is

$$Q^{(1)} = \frac{1}{A} \int_A \frac{2}{\pi} \int_{\varphi=0}^{\pi/2} \left(\frac{a_0}{a_c}\right)^{r-1} dA \, d\varphi \qquad (4.68)$$

and with the critical crack depth a_c from Equation (4.57), it is

$$Q^{(1)} = \frac{1}{A} \int_A \frac{2}{\pi} \int_{\varphi=0}^{\pi/2} \left(\frac{\sqrt{a_0}\sigma_n Y \sqrt{\pi}}{K_{\text{Ic}}}\right)^{2(r-1)} dA \, d\varphi. \qquad (4.69)$$

---

[14] It will be seen later (definition of the upper integral boundary in the biaxial stress correction factor) that it is more convenient to integrate to $\pi/2$ only and multiply by 2 instead of integrating to π.

Analogous to the uniaxial case, it is $P_{s,\text{inert}} = \exp(-M \cdot Q^{(1)})$, which finally gives the *inert failure probability of an element with a random number of randomly distributed and randomly oriented surface cracks*:

$$P_{f,\text{inert}} = 1 - \exp\left\{-\frac{1}{A_0}\int_A \frac{2}{\pi}\int_{\varphi=0}^{\pi/2} M_0 \left(\frac{\sqrt{a_0}\sigma_n Y \sqrt{\pi}}{K_{Ic}}\right)^{2(r-1)} dA\, d\varphi\right\}$$

$$= 1 - \exp\left\{-\frac{1}{A_0}\int_A \frac{2}{\pi}\int_{\varphi=0}^{\pi/2} \left(\frac{\sigma_n(\vec{r},\varphi)}{\theta_0}\right)^{m_0} dA\, d\varphi\right\} \quad (4.70)$$

4.3.4 Extension to time-dependent loading

In order to extend the failure probability equation to time-dependent loading, the time-dependent formulation of the critical crack depth $a_c(t)$ from Section 4.2.2 can be used. In the absence of subcritical crack growth, a crack fails at the point in time t if its depth a is larger than the momentary critical crack depth $a_c(t)$. The probability of an individual crack having failed up till t is thus:

$$P^{(1)}_{f,\text{inert}}(t) = \mathscr{P}\left(\exists \tau \in [0,t] : a \geq a_c(\tau)\right) = \mathscr{P}\left(a \geq \min_{\tau \in [0,t]} a_c(\tau)\right)$$

$$= 1 - \mathscr{P}\left(a < \min_{\tau \in [0,t]} a_c(\tau)\right) = 1 - F_a\left(\min_{\tau \in [0,t]} a_c(\tau)\right) \quad (4.71)$$

With the CDF of the crack depth F_a from Equation (4.36) and the parameter m_0 from Equation (4.48), this equals:

$$P^{(1)}_{f,\text{inert}}(t) = \left(\frac{a_0}{\min_{\tau \in [0,t]} a_c(\tau)}\right)^{m_0/2} \quad (4.72)$$

Following the reasoning of the preceding sections, this gives

$$P_{f,\text{inert}}(t) = 1 - \exp\left\{-\frac{1}{A_0}\int_A \frac{2}{\pi}\int_{\varphi=0}^{\pi/2} \left(\frac{\max_{\tau \in [0,t]} \sigma_n(\tau,\vec{r},\varphi)}{\theta_0}\right)^{m_0} dA\, d\varphi\right\} \quad (4.73)$$

4.3.5 Extension to account for subcritical crack growth

Subcritical crack growth (cf. Section 4.1) makes the surface flaw population time-dependent. Compared to Equation (4.71), not only the *critical*, but also the *momentary*, crack depth is now time-dependent:

$$P^{(1)}_f(t) = \mathscr{P}\left(\exists \tau \in [0,t] : a(\tau) \geq a_c(\tau)\right) \quad (4.74)$$

The criterion for the initial crack depth given in Equation (4.20) enables Equation (4.74) to be expressed as:

$$P^{(1)}_f(t) = \mathscr{P}\left(\exists \tau \in [0,t] : a_i \geq \tilde{a}_c(\tau)\right) \quad (4.75)$$

This means that instead of a criterion for the *momentary* crack depth $a(\tau)$, there is now a criterion for the *initial* crack depth a_i. One can, therefore, proceed in the same way as in Section 4.3.4, in which the crack depth is time-independent and thus always equal to a_i. Using once more the CDF of the crack depth F_a from Equation (4.36) and the parameter m_0 from Equation (4.48), it is:

$$P^{(1)}_f(t) = \mathscr{P}\left(a_i \geq \min_{\tau \in [0,t]} \tilde{a}_c(\tau)\right) = \left(\frac{a_0}{\min_{\tau \in [0,t]} \tilde{a}_c(\tau)}\right)^{m_0/2} \quad (4.76)$$

4.3. EXTENSION TO A RANDOM SURFACE FLAW POPULATION

As Equation (4.76) is completely analogous to Equation (4.72), the same procedure for generalization to an arbitrary number of randomly oriented cracks can be applied. Through considerable rearrangement but without additional simplifying assumptions, an expression for the *time-dependent failure probability of a general glass element* that takes subcritical crack growth, non-homogeneous time-variant biaxial stress fields, arbitrary geometry and arbitrary stress histories into account, can be found (see Appendix B):

$$P_{\mathrm{f}}(t) = 1 - \exp\left\{-\frac{1}{A_0}\int_A \frac{2}{\pi}\int_{\varphi=0}^{\pi/2} \max_{\tau \in [0,t]} \left\{\left[\left(\frac{\sigma_{\mathrm{n}}(\tau,\vec{r},\varphi)}{\theta_0}\right)^{n-2} + \frac{1}{U\cdot\theta_0^{n-2}}\int_0^\tau \sigma_{\mathrm{n}}^n(\tilde{\tau},\vec{r},\varphi)\,\mathrm{d}\tilde{\tau}\right]^{\frac{1}{n-2}}\right\}^{m_0} \mathrm{d}A\,\mathrm{d}\varphi\right\} \quad (4.77)$$

A_0 unit surface area, ($A_0 = 1\,\mathrm{m}^2$)

A surface area of the glass element (both faces)

t point in time

$\sigma_{\mathrm{n}}(\tau,\vec{r},\varphi)$ in-plane surface stress component normal to a crack of orientation φ at the point $\vec{r}(x,y)$ on the surface and at time τ (cf. Equation (4.62))

θ_0, m_0 surface condition parameters as defined in Equation (4.48), to be determined from experiments; $[\theta_0] = \mathrm{stress}$, $[m_0] = \mathrm{none}$

U combined coefficient containing parameters related to fracture mechanics and subcritical crack growth; $U = 2K_{\mathrm{Ic}}^2 / \left[(n-2)\cdot v_0 \cdot Y^2\pi\right]$; $[U] = \mathrm{stress}^2 \cdot \mathrm{time}$

K_{Ic} fracture toughness (cf. Section 4.2.1)

v_0, n crack velocity parameters (cf. Section 4.1.3)

Y geometry factor (cf. Section 4.2.1)

This formulation does not account for a crack growth threshold. Section 5.2.2 discusses how this can be done and whether it should be done for glass design. The only difference with respect to the inert strength formulation from Equation (4.73) is the second addend' in the integrand to account for subcritical crack growth. The discretization of Equation (4.77) for its implementation in software is explained in Section 8.5.2.

4.4 Summary and Conclusions

⇨ **Lifetime prediction model.** In this chapter, a lifetime prediction model for structural glass elements was derived based on fracture mechanics and the theory of probability. This model can be used to calculate either the time-dependent failure probability for a given action history (predictive modelling') or the maximum action history for a given service life and target failure probability (structural design). Aiming at consistency, flexibility, and a wide field of application, the lifetime prediction model offers significant advantages over currently used models:

- A comprehensive and clear derivation, which starts from the beginning, enables the model and its hypotheses to be fully understood by its users.
- The model takes all aspects that influence the lifetime of structural glass elements into consideration. It contains no simplifying hypotheses which would restrict its applicability to special cases. It is, therefore, valid for structural glass elements of arbitrary geometry and loading, including in-plane or concentrated loads, loads which cause not only time-variant stress levels but also time-variant stress distributions, stability problems, and connections.
- The parameters of the model have a clear physical meaning that is apparent to the engineer. They each include only one physical aspect and they do not depend on the experimental setup used for their determination.
- The condition of the glass surface can be modelled using either a single surface flaw (SSF) or a random surface flaw population (RSFP), i. e. a large number of flaws of random depth, location and orientation. The properties of these surface condition models are independent parameters that the user can modify. This is a major advantage, especially when hazard scenarios that involve surface damage must be analysed or when data from quality control measures or research are available. (Further investigations on when to use which of the surface condition models are required and will be presented in Section 5.1 and Chapter 8).
- The model does not suffer from the two problems of current glass design methods that have given rise to fundamental doubts regarding advanced glass modelling (cf. Section 3.3.4): The material strength rightly converges on the inert strength for very short loading times or slow crack velocity and the momentary failure probability is not independent of the momentary load.

⇨ **Use of the model.** By virtue of these features, the model is suitable for investigating the issues identified in Section 3.3. It can in particular be used to:

- address the reservations commonly advanced about fracture mechanics based glass modelling;
- assess the importance of various influences about which it is unclear whether or when they need to be considered;
- identify the implicit assumptions behind current glass design methods;
- analyse glass-specific hazard scenarios in which the main danger for a structural element arises not from the load intensity but from surface damage;
- predict the lifetime of structural glass elements of arbitrary geometry and loading;
- gain better insight into how testing should be conducted in order to obtain meaningful and safe results.

⇨ **Model hypotheses and simplification.** This chapter was dedicated to the derivation of the model only. Its underlying hypotheses, as well as possibilities for the model's simplification, are discussed in Chapter 5.

Chapter

5

Discussion and Simplification of the Model

In this chapter, the main hypotheses of the model from Chapter 4 are assessed and possibilities for the model's simplification are derived. These investigations provide answers to fundamental issues such as the relevance of the risk integral and the necessity to take the crack growth threshold and the effects of biaxial stress fields into account. Furthermore, this chapter identifies the implicit assumptions behind semi-empirical and approximate glass design methods and discusses what their range of validity is and how their parameters can (under certain conditions) be converted to the lifetime prediction model's set of fundamental parameters.

5.1 Assessment of the main hypotheses behind the model

5.1.1 Introduction

Like any engineering model, the lifetime prediction model is only useful if its hypotheses can be shown to be suitable and realistic. When deriving the model in Chapter 4, the surface condition of structural glass elements was first modelled using a single surface flaw (SSF). The model was then extended to a random surface flaw population (RSFP). Chapter 8 will focus on when to use which of these surface condition models. SSF-based modelling is rather straightforward. With an RSFP, however, lifetime prediction gets more complex and the model includes a set of additional hypotheses. These require a more in-depth discussion. To this end, it is first of all crucial to become aware of the hypotheses that have been made in the derivation:

1. The material contains a large number of natural flaws of variable depth.
2. The mechanical behaviour of the flaws can be modelled using linear elastic fracture mechanics.
3. The crack depth is a random variable that can be represented by a statistical (Pareto) distribution.
4. The number of flaws on the surface of glass elements follows a statistical (Poisson) distribution.
5. All crack locations and orientations have the same probability of occurrence which means that the corresponding random variables have uniform distributions.
6. The individual surface flaws do not influence each other.
7. The element fails when the first flaw fails (weakest-link model).
8. The geometry factor Y is independent of the crack depth.
9. Pure mode I crack propagation and failure represents the actual multimodal behaviour with sufficient accuracy.
10. Crack growth is accurately modelled by the v-K_I relationship given in Equation (4.14).

5.1.2 Strength data

By simplifying Equation (4.77), it is seen that inert strength data obtained in tests with a uniaxial or equibiaxial stress field should follow a two-parameter Weibull distribution. It is important to be aware of the fact that the Weibull distribution itself is *not* a hypothesis but a consequence of the other, more fundamental hypotheses 1, 2, 3, 4, 7 and 8. The validity of these hypotheses can, therefore, be verified by examining the goodness-of-fit of experimental data to the Weibull distribution.

To date the large majority of glass testing has been conducted at ambient conditions. Therefore, inert strength data are rare. Two published experimental data sets are plotted in Figure 5.1. The data were obtained from experiments at inert conditions on soda lime silica glass and on borosilicate glass (Schott 3111d) specimens. The figure shows a Weibull probability plot[1]. Graphically, both data sets provide a very satisfying fit. To further assess the goodness-of-fit numerically, the Anderson-Darling test is used. It yields a scalar value, the so-called observed significance level probability or p-value, which is a quantitative measure of the goodness-of-fit of data to a function. The test, the reasons for its selection and the numerical procedure used to calculate the p-value p_{AD} are explained in Section E.3.4. The resulting p_{AD}-values are given in Figure 5.1. They quantify the goodness-of-fit of the experimental data to the Weibull distributions (dotted lines). The distribution parameters have been estimated from the data. The p_{AD}-values lie clearly above 0.05, thus indicating a very high probability that the experimental data follows a Weibull distribution.

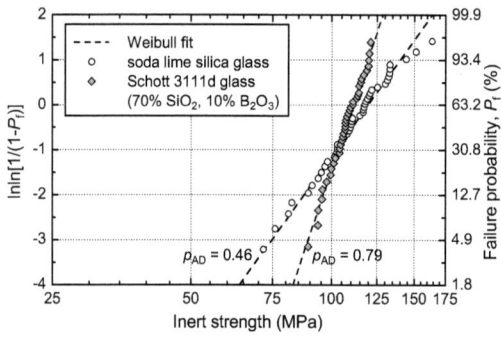

Figure 5.1:
Experimental inert strength data (cf. Section 4.2.2) of as-received glass specimens; Weibull plot. (soda lime silica glass data from Ritter et al. (1985), Schott 3111d data from Thiemeier (1989))

Again by simplification of Equation (4.77), it is seen that even strength data obtained at ambient conditions should follow a two-parameter Weibull distribution if the crack velocity parameters are constant. Experimental data, however, seem often to fit rather poorly. An example data set is shown in Figure 5.2, which is again a Weibull plot. The data points are clearly not on a straight line. Furthermore, the p_{AD}-value is 0.0000, which indicates that the hypothesis of the data following a Weibull distribution must definitely be rejected. Figure 5.3 shows again data from experiments at ambient conditions, but this time not on as-received glass specimens but on specimens with artificially induced homogeneous surface damage. It should be noted that the strength values are substantially lower compared to those in the two preceding figures. The strength scale needs, therefore, to be adapted. Failure to do so would make the goodness-of-fit look better than it actually is. It can be seen graphically, as well as from the p_{AD}-values, that the goodness-of-fit to a Weibull distribution is much better for strength data from homogeneously damaged specimens than for as-received glass data, but clearly not as good as for inert strength data.

As a consequence of the poor fit of the results of common ambient strength tests to the Weibull distribution, several researchers have proposed using log-normal or normal distributions to represent

---

[1] This is a graphical technique for determining if a data set comes from a population that fits a two-parameter Weibull distribution (Nelson, 2003). The scales of the Weibull plot are designed so that if the data follow a Weibull distribution, the data points will be on a straight line. This is why the fitted Weibull distributions, plotted as dashed lines in the figure, are straight lines.

5.1. ASSESSMENT OF THE MAIN HYPOTHESES BEHIND THE MODEL

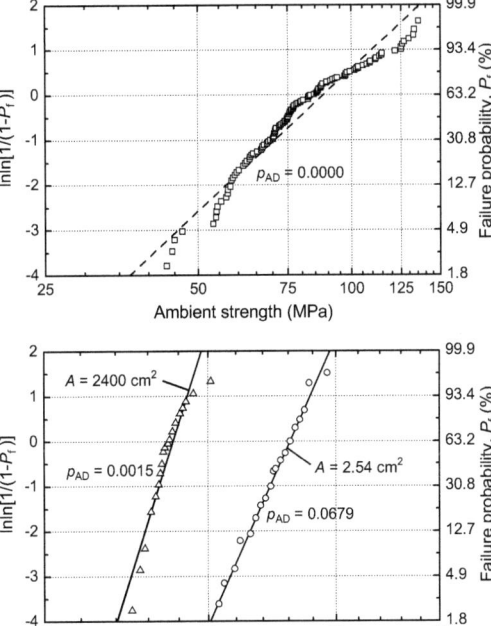

Figure 5.2: Experimental ambient strength data of as-received glass specimens; Weibull plot. (Data from Fink (2000), concentric double ring tests on as-received soda lime silica glass specimens, stress rate = 2 MPa/s, $A = 2.38 \cdot 10^{-3}$ m^2)

Figure 5.3: Experimental ambient strength data obtained from glass specimens that have been homogeneously damaged by corundum sand P16; Weibull plot. (Data from Blank (1993); 11.4 kg sand per m^2, height of fall 1.5 m, concentric double ring tests, stress rate = 2 MPa/s)

glass strength (cf. Section 3.3). This does not seem to be a satisfactory solution for the following reasons:

- As mentioned above, the Weibull distribution is a consequence of the fundamental hypotheses of the lifetime prediction model. Its replacement by another distribution type would imply that the model and its hypotheses are not valid. It is inherently inconsistent to define a characteristic strength value based on a non-Weibull distribution that is then used in combination with concepts such as the size effect (which is well confirmed in Figure 5.3) that are closely related to the hypotheses that lead to the Weibull distribution (weakest-link model).

- It was shown above that inert strength data fit very well to the Weibull distribution (Figure 5.1). The fit of ambient strength data from specimens with homogeneously damaged surfaces (Figure 5.3) is not as close. As-received glass data do not fit at all. This behaviour can be related to the influence of subcritical crack growth. At inert conditions, cracks do not grow at all, inert strength data therefore solely represent the surface condition. Strength data obtained at ambient conditions are additionally influenced by subcritical crack growth. The influence increases as time to failure increases. The influence of subcritical crack growth on strength is thus much greater with as-received specimens than with homogeneously damaged specimens. This means that in the data shown, the goodness-of-fit is inversely proportional to the influence of subcritical crack growth. This suggests that the problem is not the RSFP-based lifetime prediction model, but crack growth behaviour. The crack velocity parameters are likely to vary and thus to have a barely predictable effect on ambient glass strength.

Hypothesis 10 has been sufficiently validated over the last few decades, see Section 4.1. It must, however, be verified whether it is possible to determine the parameters with sufficient accuracy and whether they are reasonably constant in common structural applications (cf. discussion above). This is done in Chapter 6. Hypotheses 6 and 9 were discussed in Chapter 4. It was found there that hypothesis 6 is a conservative assumption and hypothesis 9 is a reasonable choice that has been confirmed by experiments. Hypothesis 5 is also a reasonable choice if the surface condition is well represented by a random surface flaw population. While this is the case for as-received glass and glass with artificially induced homogeneous surface damage, things may look different in other cases. This is investigated in the next section.

5.1.3 In-service conditions

Representing the surface condition by a random surface flaw population involves two hypotheses that may not necessarily be fulfilled in in-service conditions:

1. The random depth of surface cracks can be represented by a Pareto distribution.
2. All crack locations and orientations have the same probability of occurrence. Therefore, they can be modelled by uniform distributions.

Figure 5.4 illustrates the Pareto distribution of flaw depths. It implies that there are many small flaws but only a very few deep flaws (cf. Section 4.3.2). The probability that an element contains a deep flaw decreases as surface area decreases. The meaning of the second hypothesis is even more straightforward: It is assumed that there is no preferred flaw orientation and that the flaws are evenly distributed on the glass surface. If both hypotheses are fulfilled, the inert strength of glass specimens follows a two-parameter Weibull distribution (cf. Section 5.1.2).

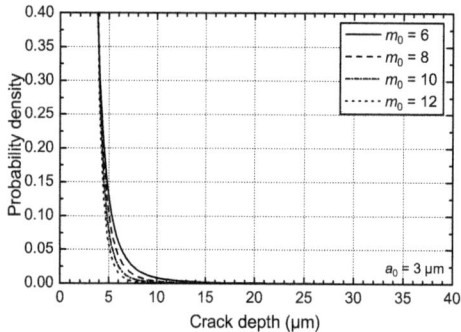

Figure 5.4:
Pareto probability density function of crack depth.

Flaw depth

If surface damage is truly random, there is a priori no reason why flaw depths should not follow a Pareto distribution. In Section 5.1, this was confirmed by experimental data. In in-service conditions, however, surface damage may not always be truly random. Calderone (2001) presented results from experiments conducted on rectangular window panels. The sample contained annealed glass from a building which was over 30 years old and new annealed glass. The panel geometry ranged from 400 mm × 2000 mm to 3000 mm × 2000 mm in size and covered aspect ratios from 1 to 5. While the experimental data from different panel sizes and aspect ratios are not directly comparable because of the influence of the stress distribution within the panel, they should definitely provide evidence of a clear tendency towards higher resistances for smaller panel sizes. Figure 5.5, however, which is based on this experimental data, shows little or no correlation between the panel area and the equivalent breakage stress at the failure origin. This supports the alternative hypothesis that glass often fails because of deep flaws that are not well represented by a random surface flaw population.

Caution is required with very small specimens. As a consequence of the Pareto flaw depth approach, very small surfaces are expected to contain flaws of extremely small depth only. Even such small surfaces, however, are exposed to some handling damage. They may, therefore, contain flaws that are substantially larger than those the statistical model suggests. This is indeed observed in the experimental data from Ritter et al. (1984), see Section 6.4.

5.1. ASSESSMENT OF THE MAIN HYPOTHESES BEHIND THE MODEL

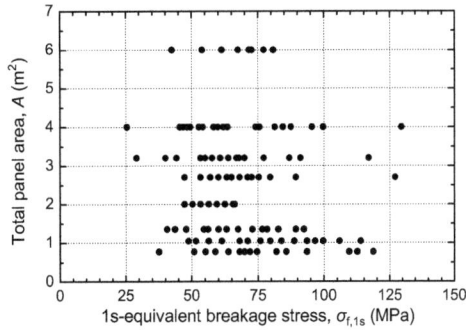

Figure 5.5:
Equivalent breakage stress (at the failure origin) of annealed glass panels as a function of the panel size (experimental data from Calderone (2001), as-received and weathered window panels).

Flaw orientation

If an equibiaxial stress field is assumed, as is recommended for structural design (cf. Section 8.4.1), results are conservative whatever the real flaw orientation may be.

If any other stress field is used, however, it is important to verify that there is no preferred flaw orientation. The vast majority of machining flaws on the edge of a glass beam, for instance, may be oriented perpendicularly to the beam's axis. In this case, a model that combines a uniaxial stress field with a random flaw orientation would give unsafe results (cf. Section 5.2.3).

Flaw location

Again, this is of major importance with any kind of machining damage. In a structural glass element containing bolt holes, for instance, the boring process is likely to cause large flaws. These flaws are not distributed equally over the element's surface, but are concentrated around the hole edge. This kind of damage is not well represented by a model that uses surface condition data from tests on glass surfaces. But even the use of a surface condition model that accounts for the machining damage does not solve the modelling problem. Stress concentrations around bolt holes affect only a small percentage of the element's surface. According to a model based on a uniform distribution of flaw locations, the probability of a deep flaw coinciding with a high stress value would, therefore, be very small. However, in the instance of a bolted connection in glass, deep flaws are very likely to coincide with stress concentrations. Bolt holes therefore cause a strong correlation between the two aspects and make the flaw location non-random. In such cases, assuming a random surface flaw population yields unsafe results.

5.1.4 Conclusions

⇨ According to the lifetime prediction model, the strength of a glass specimen in which the surface condition can be represented by a random surface flaw population should follow a two-parameter Weibull distribution. This has indeed been confirmed by experimental data of as-received glass specimens obtained at inert conditions.

⇨ In contrast to inert strength data, strength data obtained at ambient conditions do not solely represent a glass specimen's surface condition, but are additionally influenced by subcritical crack growth. The goodness-of-fit of ambient strength data to the Weibull distribution is found to decrease as time to failure increases, which means as the influence of subcritical crack growth increases. This suggests that the crack velocity parameters vary, thus making the effect of subcritical crack growth on the ambient strength of glass difficult to predict. Further investigations on this issue are required and will be presented in Chapter 6. If the variance in the crack velocity parameters is confirmed, the surface condition parameters (θ_0, m_0) need to be determined from experiments at inert conditions if they are to be reliable and safe.

⇨ The random surface flaw population represents the surface condition of as-received glass well. It may, however, often be unrepresentative of in-service conditions, especially if deep surface flaws occur or if structural elements contain machining damage.

5.2 Assessment of the relevance of some fundamental aspects

5.2.1 The risk integral

It was mentioned in Section 3.3.4 that fundamental doubts about current glass design methods must be allayed before the methods can be extended or even used with confidence. Two (related) problems associated with the use of the risk integral were identified:

- The risk integral is a function of the stress history only. This means that the failure probability at a point in time, which is a function of the risk integral, is actually independent of the momentary load at that time.

- The risk integral approaches zero when the loading time approaches zero. For very short loading times (or very slow subcritical crack growth), the material resistance obtained from an equivalent stress based model converges on infinity. This does not make sense. It should converge on the inert strength, which is an upper resistance limit (cf. Section 4.2.2).

Equation (4.77) addresses these problems by accounting for the fact that the crack depth at a point in time t is actually the sum of the initial crack depth and of the crack growth that occurred until time t:[2]

1. The initial depth of cracks depends only on the *initial surface condition*. It is influenced neither by the loading history nor by the crack growth behaviour. At inert conditions, in which cracks do not grow, the initial crack depth and the momentary load are the only aspects that influence the failure probability.

2. The *crack growth* that occurs until time t (*damage accumulation*) is a function of the loading history and of the crack growth behaviour. Its influence increases as stress levels increase, as load durations increase and as crack growth accelerates.

The two aspects are accounted for by the addends in the max-function in Equation (4.77) as follows:

$$\underbrace{\left(\frac{\sigma_n(\tau,\vec{r},\varphi)}{\theta_0}\right)^{n-2}}_{\text{initial surface condition}} + \underbrace{\frac{1}{U \cdot \theta_0^{n-2}} \int_0^\tau (\sigma_n(\tilde{\tau},\vec{r},\varphi))^n \, d\tilde{\tau}}_{\text{crack growth (damage accumulation)}} \tag{5.1}$$

Equation (4.77) can be used to predict the resistance of glass elements even for very short loading times or for conditions with slow subcritical crack growth. In contrast to resistances obtained with current glass design methods, the resistance obtained using Equation (4.77) converges on the inert strength and not on infinity.

This shows that the problem with risk integral-based glass failure modelling (the equivalent static stress approach) lies in the fact that this approach only accounts for damage accumulation. If, however, it can be shown that the addend related to the initial surface condition is small and therefore unimportant, risk integral based modelling would be very useful as an approximation.

From the various possible approaches to evaluating the relevance of this addend, the most simple and straightforward is pursued below. For constant stress rate $\dot\sigma$ and an equibiaxial stress field at any

---

[2] While it is currently not used with glass, this concept is known from its use with other construction materials. In steel structures, for instance, it is quite common that short loads of high intensity cause flaws to fail without substantial prior crack growth. (Example: Welding defects in a steel structure in which fatigue loads cause only low stress levels may not grow substantially because of fatigue loads, but may fail during a storm.)

5.2. ASSESSMENT OF THE RELEVANCE OF SOME FUNDAMENTAL ASPECTS

given time t with $\sigma_n(t) = \sigma_1(t) = \sigma_2(t) = \dot{\sigma} \cdot t$, Equation (4.77) can be reformulated to give the failure probability as a function of the failure stress $\sigma_f = \dot{\sigma} \cdot t_f$:

$$P_f(\sigma_f) = 1 - \exp\left\{-\frac{A}{A_0} \cdot \left[\left(\left(\frac{\sigma_f}{\theta_0}\right)^{n-2} + \frac{1}{U \cdot (n+1) \cdot \theta_0^{n-2}} \frac{\sigma_f^{n+1}}{\dot{\sigma}}\right)^{\frac{m_0}{n-2}}\right]\right\} \quad (5.2)$$

If only damage accumulation is to be taken into account, this simplifies to:

$$P_{f,\text{SCGonly}}(\sigma_f) = 1 - \exp\left\{-\frac{A}{A_0} \cdot \left[\left(\frac{1}{U \cdot (n+1) \cdot \theta_0^{n-2}} \frac{\sigma_f^{n+1}}{\dot{\sigma}}\right)^{\frac{m_0}{n-2}}\right]\right\} \quad (5.3)$$

Figure 5.6 compares these two formulations for various stress rates and two different crack velocity parameters v_0. All relevant parameters are indicated in the figure. The following can be observed:

- In the case of fast subcritical crack growth (left part of the figure), the results of the simplified formulation (Equation (5.3)) are identical to those of the exact formulation (Equation (5.2)) up to stress rates of about 100 MPa/s. At 1 000 MPa/s, the results start to differ. At 10 000 MPa/s, the simplified formulation grossly underestimates the probability of failure and leads to unrealistic and unsafe results (resistance above the inert strength).

- In the case of slow subcritical crack growth (right part of the figure), the behaviour is similar, but with results diverging at stress rates as low as 10 MPa/s. For stress rates in the order of magnitude of 100 MPa/s and above, the simplified formulation yields unusable results (resistance above the inert strength.)

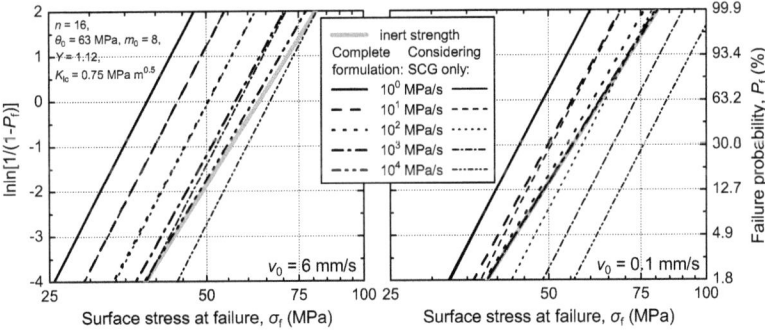

Figure 5.6: Assessing the relevance of the risk integral by comparing Equation (5.2) with Equation (5.3) for fast (left) and slow (right) crack growth.

This means that the simplified formulation in Equation (5.3), and with it the risk integral approach, is not fundamentally flawed (cf. Section 3.3.4). It is a good approximation in the case of moderate loading rates and relatively fast crack growth, i.e. when the crack depth at failure is substantially greater than the initial crack depth. This condition is met in many cases of practical relevance, in particular in structural design. For high loading rates or low crack velocities, i.e. for cases in which very little crack growth occurs, however, the simplified approach grossly underestimates the probability of failure and leads to unrealistic (resistance above the inert strength) and unsafe results. For these cases, which include the interpretation of experiments, the full formulation must be used.

In practice, neglecting the inert strength term has important advantages:

- There is no need to search for the maximum in Equation (4.77) because the addend containing the risk integral is monotonously increasing. It always reaches its maximum at $\tau = t$.

- The concept of equivalent stresses (Section 4.2.5) can be used, which simplifies calculations.

It is important to note that the crack velocity parameters n and v_0 were assumed to be constant in the present paragraph. It will be seen in Chapter 6, however, that they depend on the stress rate. Cracks grow slower as stress rates increase. Therefore, crack velocity parameters determined at low stress rates yield conservative results when used with Equation (5.2).

5.2.2 The crack growth threshold

Design for structural safety

As explained in Section 4.1.5, no subcritical crack growth occurs if the stress intensity factor K_I lies below the crack growth threshold K_{th}. Most design methods, as well as the lifetime prediction model in Equation (4.77), do not account for this. Fischer-Cripps and Collins (1995) and Overend (2002) propose a very simple and straightforward approach to account for the threshold. Its drawback is that it is valid for monotonously increasing loads only.

The problem with general, time-variant loading is that K_I lies above K_{th} during some periods, and below K_{th} during others. Because K_I depends on the stress *and* the crack depth, not only the momentary stress but also the momentary crack depth is required to determine whether there is crack growth at some point on the surface $\vec{r}(x,y)$ at a point in time t. The entire stress history $\sigma(\tau, \vec{r})$ with $\tau \in [0, t]$ must be accounted for, which results in an iterative calculation over a large number of time steps and surface sub-elements. This calculation is complex, computing time intensive and generally not worthwhile for structural applications for the following reasons:

- Neglecting the crack growth threshold is conservative (safe) for design.

- Because of the high exponent n in the crack growth law, the contribution of time periods with stress intensities below the threshold to the total crack growth over an element's lifetime, and therefore the error caused by neglecting the threshold, is generally very small. The only situation in which neglecting the threshold results in a considerable underestimation of the resistance is the rare case of a glass element that is not exposed to any high stresses during its entire design life.

Interpretation of laboratory tests

When determining surface condition parameters from laboratory tests at ambient conditions, things are different. The crack growth threshold cannot a priori be neglected because doing so leads to an overestimation of the crack growth during the tests and consequently to an underestimation of the initial flaw depths. The resulting surface condition parameters are, therefore, too optimistic and using them would be unsafe.

In order to find out whether this effect is negligible or not, the influence of the crack growth threshold in laboratory testing needs to be quantified. As an example of a common test setup, coaxial double ring tests at ambient conditions are considered. In such tests, there is a constant stress rate $\dot{\sigma}$ and an equibiaxial stress field with $\sigma_1(t) = \sigma_2(t) = \sigma_n(t) = \dot{\sigma} \cdot t$.

The probability of a crack having failed by time t is the probability that its initial depth a_i is greater than or equal to the depth $\hat{a}$ of a crack that would grow to the critical depth a_c within t. The initial depth can be found by calculating backwards from the crack depth at failure. The depth $a(t)$ of a crack that fails spontaneously at time t is:

$$a(t) = a_c(t) = \left(\frac{K_{Ic}}{\sigma_n(t) \cdot Y \sqrt{\pi}}\right)^2 = \left(\frac{K_{Ic}}{\dot{\sigma} t \cdot Y \sqrt{\pi}}\right)^2 \qquad (5.4)$$

5.2. ASSESSMENT OF THE RELEVANCE OF SOME FUNDAMENTAL ASPECTS

For monotonously increasing stress, crack growth starts at a given point in time $\hat{t}$ and continues until failure. A crack starting to grow with depth $\hat{a}$ at time $\hat{t}$ will grow to

$$a(t) = \left[\hat{a}^{\frac{2-n}{2}} + \frac{2-n}{2} \cdot v_0 \cdot K_{Ic}^{-n} \cdot \left(Y\sqrt{\pi}\right)^n \cdot \int_{\hat{t}}^{t} [\sigma(\tau)]^n \, \mathrm{d}\tau \right]^{\frac{2}{2-n}} \quad (5.5)$$

until t (see Equation (4.16)). From these two equations, $\hat{a}(t)$ is found:

$$\hat{a}(t) = \left[\left(\frac{K_{Ic}}{\sigma_n(t) \cdot Y\sqrt{\pi}} \right)^{2-n} - \frac{2-n}{2} \cdot v_0 \cdot K_{Ic}^{-n} \cdot (Y\sqrt{\pi})^n \cdot \int_{\hat{t}}^{t} (\sigma_n(\tau))^n \, \mathrm{d}\tau \right]^{\frac{2}{2-n}} \quad (5.6)$$

This is the crack depth that would cause failure after exactly t. With the Pareto distribution from Equation (4.36) ($m_0 = 2(r-1)$), the probability of a crack having failed by time t is:

$$P^{(1)}(t) = \mathscr{P}\left(a_i > \hat{a}(t)\right) = 1 - F_a\left(\hat{a}(t)\right) = \left(\frac{a_0}{\hat{a}(t)}\right)^{m_0/2} \quad (5.7)$$

Apart from the lower integral boundary $\hat{t}$, the formulation is identical to the case without a threshold. The failure probability equation can therefore easily be adapted accordingly. Considering the crack growth's contribution to the failure probability only (cf. Section 5.2.1) but taking the crack growth threshold into account, the failure probability of a specimen exposed to a monotonously increasing tensile stress equals:

$$P_f(t) = 1 - \exp\left\{ -\frac{A}{A_0} \left[\left(\frac{1}{U \cdot \theta_0^{n-2}} \int_{\hat{t}}^{t} (\sigma_n(\tau))^n \, \mathrm{d}\tau \right)^{\frac{m_0}{n-2}} \right] \right\} \quad (5.8)$$

For a constant stress rate ($\sigma(\tau) = \dot{\sigma}\tau$) and as a function of the failure stress $\sigma_f = \dot{\sigma}t$, this reads:

$$P_f(\sigma_f) = 1 - \exp\left\{ -\frac{A}{A_0} \left(\frac{1}{U \cdot \theta_0^{n-2}} \cdot \dot{\sigma}^n \cdot \frac{(\sigma_f/\dot{\sigma})^{n+1} - \hat{t}^{n+1}}{n+1} \right)^{\frac{m_0}{n-2}} \right\} \quad (5.9)$$

The only remaining issue is to find the point in time $\hat{t}$ that crack growth starts at. With the threshold stress intensity factor K_{th}, the stress-dependent and therefore time-dependent threshold crack depth is:

$$a_{th}(\tau) = \left(\frac{K_{th}}{\sigma_n(\tau) \cdot Y\sqrt{\pi}} \right)^2 = \left(\frac{K_{th}}{\dot{\sigma}\tau \cdot Y\sqrt{\pi}} \right)^2 \quad (5.10)$$

From the introductory definitions it is known that $\hat{a}(t) = a_{th}(\hat{t})$. This enables $\hat{t}$ to be found numerically (still for a constant stress rate coaxial double ring test setup):

$$f_{\hat{t}}(\hat{t}) \stackrel{!}{=} 0$$

$$f_{\hat{t}}(\hat{t}) = \left(\frac{K_{Ic}}{\dot{\sigma}t \cdot Y\sqrt{\pi}} \right)^{2-n} - \left(\frac{K_{th}}{\dot{\sigma}\hat{t} \cdot Y\sqrt{\pi}} \right)^{2-n} - \frac{2-n}{2} \cdot v_0 \cdot \left(\frac{Y\sqrt{\pi}}{K_{Ic}} \right)^n \cdot \dot{\sigma}^n \cdot \frac{t^{n+1} - \hat{t}^{n+1}}{n+1} \quad (5.11)$$

Using this approach, the influence of the crack growth threshold K_{th} in coaxial double ring tests is visualized in Figure 5.7. The four sub-figures show the results for two different constant stress rates and crack velocity parameters as indicated. It can be seen that the influence of the stress rate decreases as stress rates increase and crack velocities decrease. There is no point in performing coaxial double ring tests at stress rates below 0.02 MPa/s and linear crack velocity parameters above 6 mm/s are very unlikely in laboratory tests (cf. Section 6.3). Therefore, the upper right sub-figure represents a worst case scenario. In view of this and the fact that K_{th}/K_{Ic} is about 0.3 for soda lime silica glass (see Section 6.2), it can be concluded that the crack growth threshold of soda lime silica glass can safely be neglected in common constant stress rate tests.

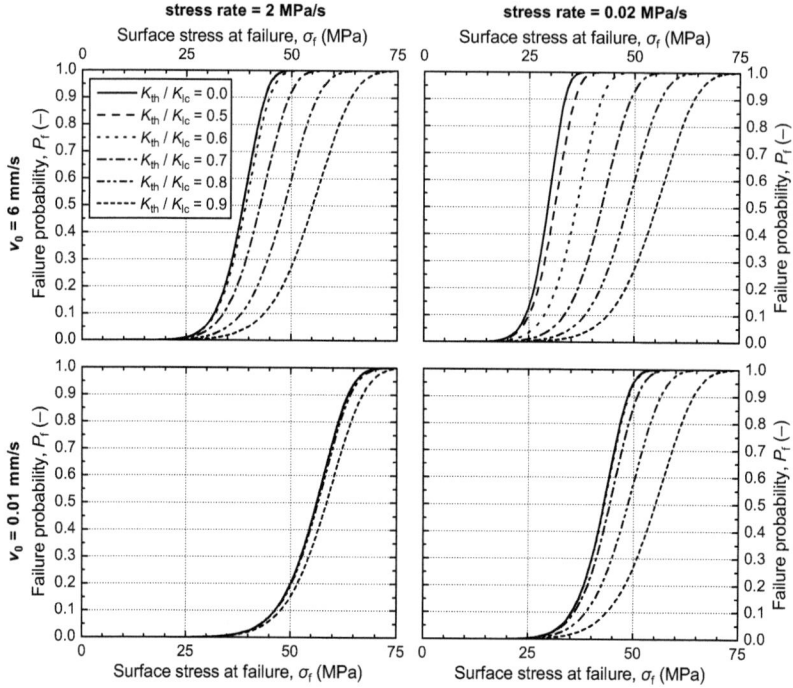

Figure 5.7: The influence of the crack growth threshold K_{th} in common coaxial double ring tests at ambient conditions. The constant stress rate and the crack velocity parameter v_0 vary between figures as indicated. All other parameters, namely $Y = 1.12$, $K_{\text{Ic}} = 0.75\,\text{MPa}\,\text{m}^{0.5}$, $n = 16$, $\theta_0 = 63\,\text{MPa}$, $m_0 = 8.0$ and $A = 1\,\text{m}^2$, are the same in all sub-figures.

5.2.3 Biaxial stress fields

Introduction

The lifetime prediction model enables the issues identified in Section 3.3.8 to be addressed. It has already been stated there that while assuming an equibiaxial stress field is safe for design, it is unsafe when deriving glass strength data from tests. In a four point bending test, for instance, the specimen is exposed to a *uniaxial stress field* ($\sigma_1 \neq 0$, $\sigma_2 = 0$). Only a minority of the surface flaws are oriented perpendicularly to the major principal stress and consequently exposed to high stress (Equation (4.62) with $\sigma_2 = 0$, $\varphi \approx 0$). In an *equibiaxial stress field*, however, all cracks are exposed to the stress σ_1, whatever their orientation may be. This is why coaxial double ring tests, in which specimens are exposed to an equibiaxial stress field, give lower strength values than do any other testing procedures. Consequently, the use of strength values from coaxial double ring tests for design is safe, while the use of four point bending strength values is not.

In the following, the lifetime prediction model is used to find a simplified way of accounting for a biaxial stress field. Furthermore, the range of validity of this simplification is discussed. A moderate rate of loading is assumed, such that Equation (4.77) can be simplified by considering damage

5.2. ASSESSMENT OF THE RELEVANCE OF SOME FUNDAMENTAL ASPECTS

accumulation only (cf. Section 5.2.1):

$$P_f(t) = 1 - \exp\left\{-\frac{1}{A_0}\int_A \frac{2}{\pi}\int_{\varphi=0}^{\pi/2}\left[\frac{1}{U\cdot\theta_0^{n-2}}\int_0^t (\sigma_n(\tau,\vec{r},\varphi))^n\,d\tau\right]^{\frac{m_0}{n-2}}dA\,d\varphi\right\} \quad (5.12)$$

Time-independent principal stress ratio

If the principal stress ratio $R(\vec{r}) = \sigma_2(\vec{r})/\sigma_1(\vec{r})$ remains constant for all times $\tau \in [0, t]$ and at all points on the surface $\vec{r}(x, y) \in A$, Equation (5.12) can be simplified using integral separation. With Equation (4.62) and $R(\vec{r})$, the stress component normal to a crack can be expressed as:

$$\sigma_n(\tau,\vec{r},\varphi) = \sigma_1(\tau,\vec{r})\left[\cos^2\varphi + R(\vec{r})\sin^2\varphi\right] \quad (5.13)$$

With this and the t_0-equivalent major principal stress $\sigma_{1,t_0}(t,\vec{r})$ calculated from $\int_0^t \sigma_1^n(\tau,\vec{r})\,d\tau = \sigma_{1,t_0}^n(t,\vec{r})\cdot t_0$, Equation (5.12) is

$$P_f(t) = 1 - \exp\left\{-\frac{1}{A_0}\left(\frac{t_0}{U\cdot\theta_0^{n-2}}\right)^{\frac{m_0}{n-2}}\int_A \left[\eta_b(\vec{r})\cdot\sigma_{1,t_0}(t,\vec{r})\right]^{\frac{nm_0}{n-2}}dA\right\}, \quad (5.14)$$

in which the location-dependent factor $\eta_b(\vec{r})$ is:

$$\eta_b(\vec{r}) = \left[\frac{2}{\pi}\int_{\varphi=0}^{\pi/2}\left[\cos^2\varphi + R(\vec{r})\sin^2\varphi\right]^{\frac{nm_0}{n-2}}d\varphi\right]^{\frac{n-2}{nm_0}}. \quad (5.15)$$

Integrating over the whole range of φ is correct for $\sigma_1 \geq \sigma_2 \geq 0$ only. If $\sigma_2 < 0$, some φ give negative σ_n that have no crack-opening effect. To account for this, the upper limit of the integral needs to be adjusted to integrate only up to the angle $\hat{\varphi}$, which is the angle at which σ_n becomes negative. From $\cos^2\hat{\varphi} + R(\vec{r})\sin^2\hat{\varphi} \stackrel{!}{=} 0$, this angle is found to be $\hat{\varphi} = \arctan\left[(-R)^{-1/2}\right]$. The *location-dependent biaxial stress correction factor for a time-independent principal stress ratio* is thus:

$$\eta_b(\vec{r}) = \left[\frac{2}{\pi}\cdot\int_{\varphi=0}^{\hat{\varphi}}\left[\cos^2\varphi + R(\vec{r})\sin^2\varphi\right]^{\tilde{m}}d\varphi\right]^{1/\tilde{m}} \quad (5.16)$$

$$\hat{\varphi} = \begin{cases} \pi/2 & \text{for } \sigma_2 \geq 0 \\ \arctan\left[(-R)^{-1/2}\right] & \text{for } \sigma_2 < 0 \end{cases} \qquad \tilde{m} = \begin{cases} \frac{nm_0}{(n-2)} & \text{at ambient conditions} \\ m_0 & \text{at inert conditions} \end{cases} \quad (5.17)$$

It should be noted that this biaxial stress correction factor η_b depends on the surface condition parameter m_0, on the crack velocity parameter n and on the principal stress ratio R. As m_0 increases, η_b becomes less sensitive to R and approaches 1.

The glass failure prediction model by Beason and Morgan (see Section 3.2.7), as well as its parameters and test procedures, suffers from several shortcomings and is not directly compatible with approaches based completely on fracture mechanics. Therefore, the biaxial stress correction factor that it proposes has not found acceptance in Europe (cf. Section 3.3.8). It is, however, formally equivalent to Equation (5.16) (ambient conditions), which has been derived entirely on fracture mechanical and statistical bases. This means that the biaxial stress correction factor itself is a suitable means of taking biaxial stress fields into account. It is, however, only valid if the principal stress ratio $R(\vec{r},\tau) = \sigma_2(\vec{r},\tau)/\sigma_1(\vec{r},\tau)$ remains constant for all points in time $\tau \in [0, t]$ at all points on the surface $\vec{r}(x, y) \in A$. While this condition holds true e. g. in four point bending tests, it does not in most general cases, including rectangular glass plates exposed to uniform lateral load.

Time-dependent principal stress ratio

Accounting for biaxial stress fields becomes much more complex if the principal stress ratio is time-dependent. To illustrate the added complexity, a very common situation has been chosen, namely a rectangular glass plate exposed to uniform lateral load. Figure 5.8 shows the geometry, the support conditions and the location of sections A–A and B–B used in subsequent figures.

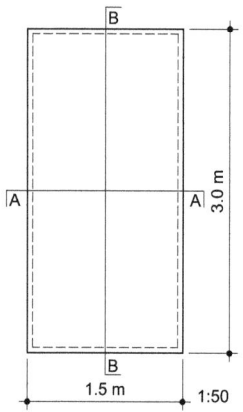

Figure 5.8:
Geometry, support conditions and sections. The dashed lines mean that the displacement perpendicular to the plate is restrained while rotation is unconstrained.

Figure 5.9 shows the minor and major in-plane principal stresses as well as the principal stress ratio R on the tension face[†] of such a plate of dimensions $3.0 \, \text{m} \times 1.5 \, \text{m}$. The different curves represent different load levels. The quantity q^* is the *normalized pressure*[3], which is defined as follows[4]:

$$q^* = \frac{qb^4}{Dh} \quad \text{with} \quad D = \frac{Eh^3}{12(1-\nu^2)} \tag{5.18}$$

The variable b is the short edge length of the plate, q is the uniform lateral load, h is the thickness of the plate, E is Young's modulus and ν is Poisson's ratio. In Figure 5.10, the resulting biaxial stress correction factor η_b is shown for sections A–A and B–B as well as for the entire plate for a very low and a very high load level. The results shown in the figure are calculated using $m_0 = 6$. The biaxial stress correction factor is not only location-dependent, but it furthermore varies considerably as load levels change. The use of a location-dependent but time-independent η_b cannot, therefore, be expected to give accurate results. On the other hand, η_b varies only between 0.8 and 1.0 within regions of substantial tensile stress. As can be seen from Figure 5.11, it further approaches 1 for higher values of m_0.

In conclusion, the following is recommended for cases with time-variant principal stress ratios:

- For *design calculations*, to assume an equibiaxial stress field ($\eta_b = 1$) is a good solution. This assumption greatly simplifies the model and it is safe without being excessively conservative.

- For the *interpretation of test data*, a transient numeric calculation[†] (see Section 8.5.2) should be performed in order to avoid unsafe results.

---

[3] The advantage of the normalized pressure is that the figures in this paragraph apply to any plate stiffness (i.e. glass thickness).

[4] Caution: Sometimes (e.g. in prEN 13474-2:2000) a different definition not including ν is used to define a quantity that is equally called 'normalized pressure': $q^{**} = qb^4/(h^4E)$.

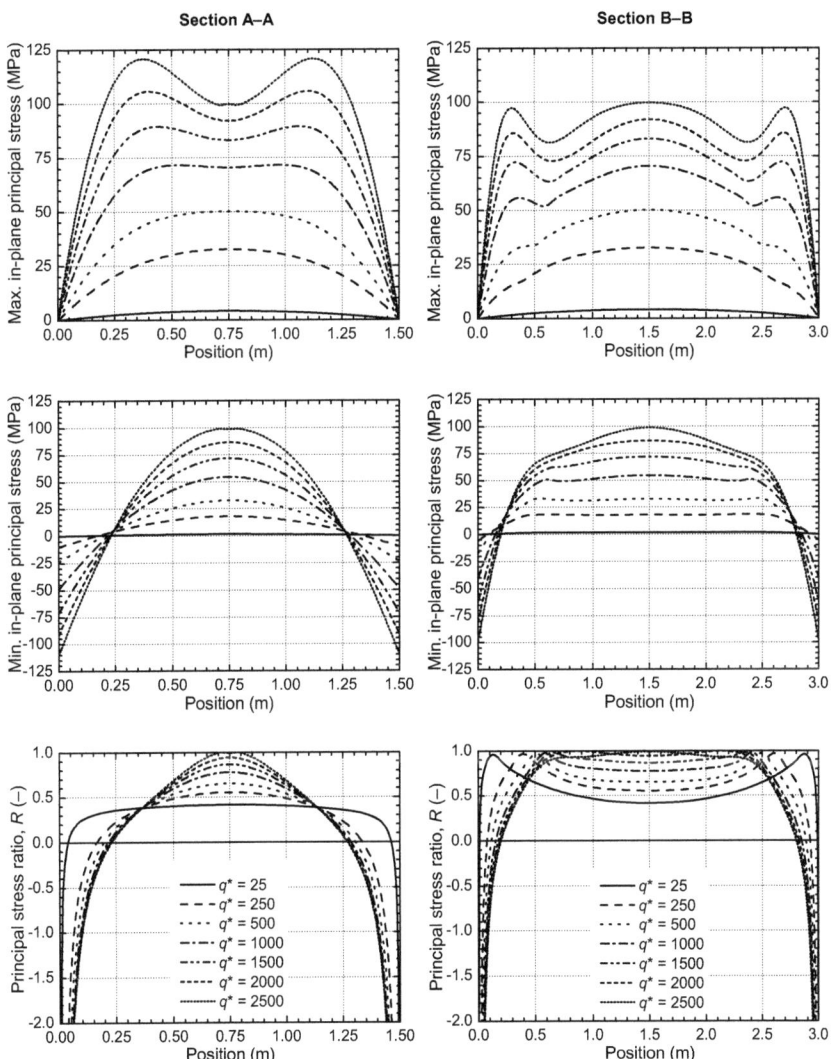

Figure 5.9: Minor and major in-plane principal stresses and principal stress ratios on the tension face of a glass plate as a function of the normalized pressure.

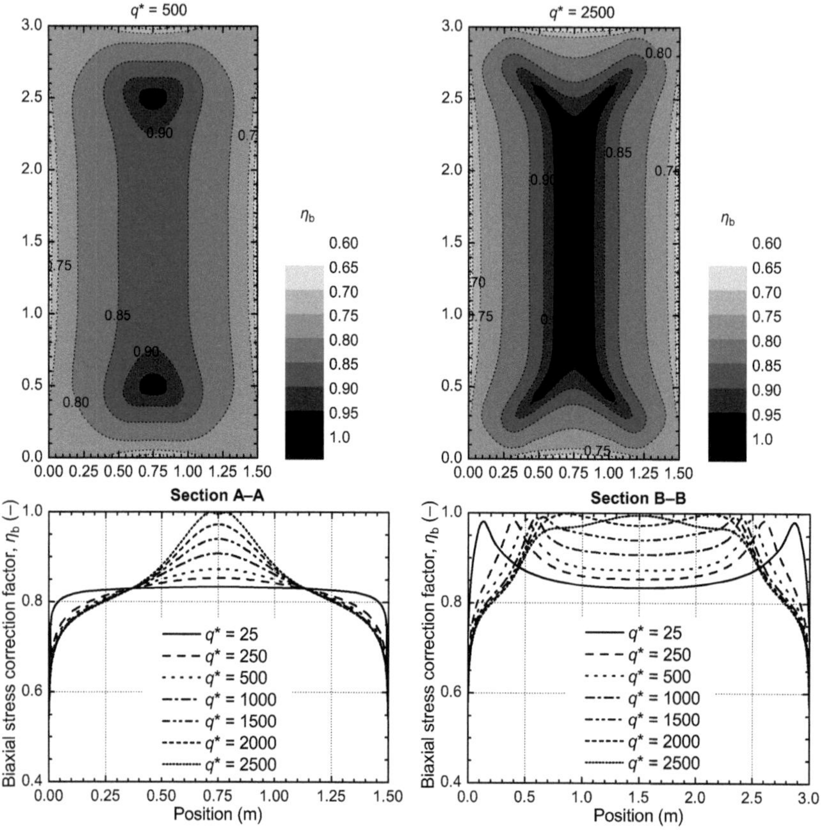

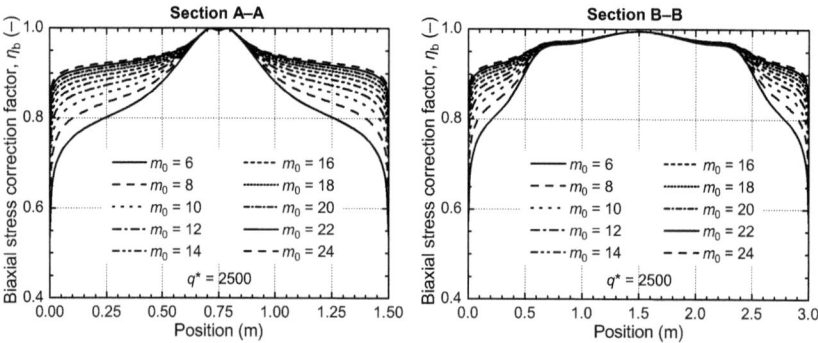

Figure 5.10: Biaxial stress correction factor in a glass plate as a function of the normalized pressure.

Figure 5.11: Biaxial stress correction factor in a glass plate as a function of the surface condition parameter m_0 and for $q^* = 2500$.

5.3 Simplification for special cases

5.3.1 Structural Design

The following simplifications are appropriate for the vast majority of common structural glass design tasks:

- Calculating the failure probability on the basis of the risk integral is an approximation of sufficient accuracy (cf. Section 5.2.1).
- The crack growth threshold can be neglected (cf. Section 5.2.2).
- An equibiaxial stress field may be assumed (cf. Section 5.2.3).

With these simplifying assumptions and using the t_0-second major principal stress

$$\sigma_{1,t_0}(t,\vec{r}) = \left(\frac{1}{t_0}\int_0^t \sigma_1^n(\tau,\vec{r})\,\mathrm{d}\tau\right)^{1/n} \approx \left(\frac{1}{t_0}\sum_{j=1}^J \left[\sigma_{1,\tau_j}^n \cdot \tau_j\right]\right)^{1/n} \tag{5.19}$$

from Section 4.2.5, Equation (4.77) simplifies to

$$P_\mathrm{f}(t) = 1 - \exp\left\{-\frac{1}{A_0}\left(\frac{t_0}{U \cdot \theta_0^{n-2}}\right)^{\frac{m_0}{n-2}}\int_A \left[\sigma_{1,t_0}(t,\vec{r})\right]^{\frac{nm_0}{n-2}} \mathrm{d}A\right\}. \tag{5.20}$$

Since it yields a failure probability, this is still a model and not a design equation. For design, the failure probability is not a result but a target value (cf. Chapter 8). This means that the target failure probability needs to be introduced as an additional parameter. Furthermore, the standard verification format that engineers are used to involves the comparison of a resistance term to an action term. Equation (5.20) needs to be reformulated accordingly. Finally, further simplification of the equation would be convenient.

The first step is to define a uniformly distributed stress $\sigma_{1\mathrm{eq},t_0}$ that would have the same effect as the actual stress distribution, namely $\int_A \sigma_{1,t_0}^{\tilde{m}}\,\mathrm{d}A = A \cdot \sigma_{1\mathrm{eq},t_0}^{\tilde{m}}$. Using the combined parameter $\tilde{m} = n\, m_0/(n-2)$ as in Equation (5.17)[5], this equivalent uniformly distributed stress is:

$$\sigma_{1\mathrm{eq},t_0} = \left(\frac{1}{A}\int_A \sigma_{1,t_0}^{\tilde{m}}\,\mathrm{d}A\right)^{1/\tilde{m}} \tag{5.21}$$

Secondly, this stress value can be further standardized by defining the equivalent stress that would have the same effect when acting on the unit surface area $A_0 = 1\,\mathrm{m}^2$ instead of A. This yields the equivalent t_0-second uniform stress on the unit surface area (in short: equivalent reference stress):

$$\tilde{\sigma} = \sigma_{1\mathrm{eq},t_0,A_0} = \left(\frac{A}{A_0}\right)^{1/\tilde{m}} \sigma_{1\mathrm{eq},t_0}$$

$$= \left(\frac{1}{A_0}\int_A \sigma_{1,t_0}^{\tilde{m}}\,\mathrm{d}A\right)^{1/\tilde{m}} \approx \left(\frac{1}{A_0}\sum_i \left[A_i \cdot \sigma_{1,t_0,i}^{\tilde{m}}\right]\right)^{1/\tilde{m}} \tag{5.22}$$

The formulation with the summation sign is used when processing finite element results, in which the total element surface A with a non-uniform stress distribution is divided into sub-surfaces of

---

[5] The relationship $\tilde{m} = m_0$ given for inert conditions is *not* valid here because the simplifying assumptions behind Equation (5.20) do not allow this equation to be used for inert conditions.

areas A_i with an approximately homogeneous stress field (cf. Section 8.5.2). Inserting this into Equation (5.20) and introducing one more combined parameter ($\bar{k}$) yields:

$$P_f(t) = 1 - \exp\left\{-\bar{k}\bar{\sigma}^{\bar{m}}\right\} \tag{5.23}$$

$$\bar{m} = \frac{n\,m_0}{(n-2)} \qquad \bar{k} = \left(\frac{t_0}{U \cdot \theta_0^{n-2}}\right)^{\frac{m_0}{n-2}} \tag{5.24}$$

Provided that the stress history of all sub-surfaces is known, $\bar{\sigma}$ can be evaluated for any conditions. Rearrangement yields a standard failure criterion:

$$\bar{\sigma} < f_0(P_{f,t}) \tag{5.25}$$

$$f_0(P_{f,t}) = \left[-\ln(1 - P_{f,t})\right]^{1/\bar{m}} \cdot \left(\frac{t_0}{U \cdot \theta_0^{n-2}}\right)^{-1/n} = f_{0,\text{inert}}^{(n-2)/n} \cdot \left(\frac{U}{t_0}\right)^{1/n} \tag{5.26}$$

The reference inert strength $f_{0,\text{inert}}$ is defined in Equation (4.52) and U is defined in Equation (4.76). The resistance term $f_0(P_{f,t})$, called *reference ambient strength* hereafter, is a function of the *target failure probability* $P_{f,t}$ and has the following physical meaning: The failure probability of a glass element with surface area $A_0 = 1\,\text{m}^2$ that is exposed to a uniformly distributed crack opening surface stress f_0 for $t_0 = 1\,\text{s}$ at ambient conditions is $P_{f,t}$.

It is important to be aware of the parameter dependencies:

- The equivalent reference stress $\bar{\sigma}$ is a function of the loading history (intensity and shape), the residual stress σ_r, the element surface area A, the exponential crack velocity parameter n and the surface condition parameter m_0.
- The reference inert strength $f_{0,\text{inert}}$ is a function of the target failure probability $P_{f,t}$ and of the surface condition parameters θ_0 and m_0 only. The reference ambient strength f_0 depends additionally on the crack velocity parameters v_0 and n. In contrast to common measures for glass resistance (cf. Section 3.2), however, it is independent of the loading history, the surface area A and the residual stress σ_r.

5.3.2 Avoiding transient finite element analyses

In Section 5.3.1, the lifetime prediction model from Equation (4.77) was simplified considerably by neglecting some aspects that are of limited influence in common design tasks. The quantification of the equivalent reference stress (Equation (5.22)), however, still requires a transient finite element analysis. This is unproblematic for research activities, for simple loading histories and for the interpretation of experimental data. For application in practice, however, such analyses are often too complex and time-consuming . Therefore, the present section discusses how and under what conditions transient finite element analyses can be avoided. This discussion also provides additional insight into the simplifying assumptions behind current glass design methods.

A time-dependent non-uniform stress field $\sigma(\tau, \vec{r})$ can be expressed in terms of a *representative stress* $\breve{\sigma}(\tau)$ at one point on the surface and a *dimensionless stress distribution function* $c(\tau, \vec{r})$:

$$\sigma(\tau, \vec{r}) = \breve{\sigma}(\tau) \cdot \left(\frac{\sigma(\tau, \vec{r})}{\breve{\sigma}(\tau)}\right) = \breve{\sigma}(\tau) \cdot c(\tau, \vec{r}) \tag{5.27}$$

The maximum stress on an element $\sigma_{\max}(\tau)$ is generally a sensible choice for the representative stress $\breve{\sigma}(\tau)$. It is important to bear in mind that $\sigma(\tau, \vec{r})$ refers to the crack opening stress (cf. Section 4.2.3), such that $c(\tau, \vec{r}) = \sigma(\tau, \vec{r})/\breve{\sigma}(\tau)$ is equal to zero in compressed regions of the surface.

The dimensionless stress distribution function allows for the rearrangement of Equation (5.22) as follows:

$$\bar{\sigma} = \left(\frac{1}{A_0}\int_A \left[\left(\frac{1}{t_0}\cdot\int_0^t (\breve{\sigma}(\tau)\cdot c(\tau,\vec{r}))^n\,d\tau\right)^{\bar{m}/n}\right]\,dA\right)^{1/\bar{m}} \tag{5.28}$$

5.3. SIMPLIFICATION FOR SPECIAL CASES

If $c(\tau, \vec{r})$ is independent of the load level represented by the time-dependent representative stress $\check{\sigma}(\tau)$ and therefore independent of the time τ, it can be isolated from the time-integral:

$$\bar{\sigma} = \left(\frac{1}{A_0} \int_A \left[c^{\tilde{m}}(\vec{r}) \left(\frac{1}{t_0} \int_0^t \check{\sigma}^n(\tau) \, d\tau \right)^{\tilde{m}/n} \right] dA \right)^{1/\tilde{m}} \tag{5.29}$$

This allows the time-integral and the area-integral to be separated

$$\bar{\sigma} = \left(\frac{1}{A_0} \right)^{1/\tilde{m}} \cdot \left(\frac{1}{t_0} \int_0^t \check{\sigma}^n(\tau) \, d\tau \right)^{1/n} \cdot \left(\int_A c(\vec{r})^{\tilde{m}} \, dA \right)^{1/\tilde{m}} \tag{5.30}$$

such that the equivalent reference stress $\bar{\sigma}$ can be expressed as follows:

$$\boxed{\bar{\sigma} = A_0^{-1/\tilde{m}} \cdot \check{\sigma}_{t_0} \cdot \bar{A}^{1/\tilde{m}} \quad \text{with} \quad \bar{A} = \int_A c(\vec{r})^{\tilde{m}} \, dA \approx \sum_i A_i c_i^{\tilde{m}}} \tag{5.31}$$

$\check{\sigma}_{t_0}$ is the t_0-second *equivalent representative stress* (calculated from $\check{\sigma}(\tau)$ using Equation (4.22)). The *equivalent area* $\bar{A}$ (also known as the *effective area*) is the surface area of a glass element that fails with the same probability, when exposed to the uniform representative stress $\bar{\sigma}$, as an element with surface area A fails when exposed to the non-uniform stress field $\sigma(\vec{r})$. $\bar{A}$ can be defined for ambient and inert conditions alike, with $\tilde{m}$ being $n m_0/(n-2)$ and m_0 respectively (cf. Equation (5.17)).

The formulation in Equation (5.31) is of particular interest: It enables, for instance, convenient design aids to be created in order to avoid transient finite element analyses for common design tasks.

In fact, all current glass design methods assume Equation (5.31) to be valid, without declaring this assumption or discussing the conditions required for its validity. It should, however, be noted that Equation (5.31) is *only* valid if $\bar{A}$ and therefore the stress distribution function $c(\tau, \vec{r})$ are constant for all $\tau \in [0, t]$.

Conclusions

⇨ **General conditions.** Geometric non-linearity (e. g. because plates undergo deformations larger than their thickness), residual stress (σ_r), external constraints (σ_p), and actions that vary not only in intensity but also in shape make the dimensionless stress distribution function c and therefore the equivalent area $\bar{A}$ depend on the representative stress and therefore on time. Equation (5.31) is *not* valid in these conditions and there is no simple way to superimpose loads or to consider load duration effects. Therefore, Equation (5.22) must be solved.

⇨ **Conditions in which no transient analysis is required.** Equation (5.31) allows transient analyses to be avoided if $\bar{A}$ and therefore the stress distribution function $c(\tau, \vec{r})$ are constant for all $\tau \in [0, t]$. This requires the following two conditions to be satisfied:

1. *The decompressed surface area remains constant.* The dimensionless stress distribution function $c(\tau, \vec{r})$ has a discontinuity: it is equal to zero in compressed regions of the surface, that is where the crack opening stress $\sigma(\tau, \vec{r}) = \sigma_E(\tau, \vec{r}) + \sigma_r + \sigma_p$ (cf. Section 4.2.3) is compressive (< 0). This discontinuity does *not* occur
   a) if $\sigma(\tau, \vec{r}) > 0 \, \forall (\tau, \vec{r})$, for instance on the tension face of annealed glass panes ($\sigma_r \approx 0$) exposed to a uniform lateral load;
   b) if $\sigma(\tau, \vec{r}) \leq 0 \, \forall (\tau, \vec{r})$. This case is of no interest in the current discussion because it does not cause subcritical crack growth. (Typical examples are heat treated glass elements that are designed such that $\max_{\tau \in [0,T]} \sigma_E(\tau) \leq -(\sigma_r + \sigma_p) \, \forall \, \vec{r}$; cf. Section 4.2.3).

2. *The major principal stress is proportional to the load at all points on the surface.* This requires in particular

a) linear elastic material behaviour,
b) no (or negligible) geometrically non-linear behaviour,
c) variation of the applied load intensity only (but not of the load shape).

Under these conditions, $c(\tau, \vec{r})$ and $\bar{A}$ do *not* depend on the representative stress $\check{\sigma}(\tau)$ and are therefore time-independent. They depend solely on the shape of the stress distribution within the element and on the element's size, which are in turn both constant for the conditions at hand. Although there are many cases in which the conditions are not satisfied, current glass design methods implicitly assume that they are.

The common approach of applying the load duration effect found for a single crack (see Section 4.2) to the representative stress $\check{\sigma}$ is valid under these special conditions:

$$\frac{\check{\sigma}^{(2)}}{\check{\sigma}^{(1)}} = \left(\frac{T_1}{T_2}\right)^{1/n} \qquad (5.32)$$

From Equation (5.31) the ratio between the equivalent reference stresses $\bar{\sigma}$ of two load histories characterized by the t_0-second equivalent representative stresses $\check{\sigma}_{t_0}$ can be derived:

$$\frac{\bar{\sigma}^{(1)}}{\bar{\sigma}^{(2)}} = \frac{\check{\sigma}_{t_0}^{(1)}}{\check{\sigma}_{t_0}^{(2)}} \qquad (5.33)$$

Therefore, if the equivalent stress $\bar{\sigma}^{(1)}$ in a given loading situation of duration T_1 is known, the equivalent stress $\bar{\sigma}^{(2)}$ resulting from the same loading being applied for T_2 is:

$$\frac{\bar{\sigma}^{(2)}}{\bar{\sigma}^{(1)}} = \left(\frac{T_2}{T_1}\right)^{1/n} \qquad (5.34)$$

As two elements with $\bar{\sigma}^{(1)} = \bar{\sigma}^{(2)}$ have equal probabilities of failure, Equation (5.31) provides a simple approach to design. The t_0-second resistance of specific elements in specific conditions can be provided for instance in design tables or graphs. A structural safety verification simply involves comparing this resistance to a t_0-second equivalent representative stress $\check{\sigma}_{t_0}$ calculated from an arbitrary stress history. As the representative stress is proportional to the load ($\check{\sigma}(t) = \varepsilon \cdot q(t)$) in conditions in which $\bar{A}$ is constant, this approach can even be extended to loads. Analogous to Equation (4.22), it is:

$$\mathbb{X}_{t_0} = \left(\frac{1}{t_0}\int_0^T \mathbb{X}^n(\tau)\,d\tau\right)^{1/n} \approx \left(\frac{1}{t_0}\sum_{j=1}^J \left[\mathbb{X}_{t_j}^n \cdot t_j\right]\right)^{1/n} \quad \text{with } \mathbb{X} = \begin{cases} \check{\sigma} & \text{for stresses} \\ q & \text{for loads} \end{cases} \qquad (5.35)$$

▷ **Constant load and non-linear behaviour.** If a constant load is applied during T, $\bar{A}$ is also constant. It is

$$\check{\sigma}_{t_0} = \check{\sigma}\left(T/t_0\right)^{1/n} = \check{\sigma}\bar{t} \qquad (5.36)$$

and Equation (5.31) is applicable.

▷ **Comparing two periods of constant load on an element with non-linear behaviour.** For any two periods of duration T_1 and T_2 with constant representative stresses $\check{\sigma}^{(1)}$ and $\check{\sigma}^{(2)}$, an equal probability of failure is obtained if they give the same equivalent reference stress $\bar{\sigma}$:

$$\bar{\sigma}^{(1)} \stackrel{!}{=} \bar{\sigma}^{(2)} \quad \Longrightarrow \quad A_0^{-1/\tilde{m}} \cdot \check{\sigma}^{(1)} \bar{t}_1 \cdot \bar{A}_1^{1/\tilde{m}} \stackrel{!}{=} A_0^{-1/\tilde{m}} \cdot \check{\sigma}^{(2)} \bar{t}_2 \cdot \bar{A}_2^{1/\tilde{m}} \qquad (5.37)$$

Rearrangement and insertion of $\bar{t} = (T/t_0)^{1/n}$ yields:

$$\frac{\check{\sigma}^{(2)}}{\check{\sigma}^{(1)}} = \left(\frac{T_1}{T_2}\right)^{1/n} \left(\frac{\bar{A}_1}{\bar{A}_2}\right)^{1/\tilde{m}} \qquad (5.38)$$

This means that the 'tensile strength ratio', the ratio of the maximum allowable stress on a glass element for two periods of constant load, depends on the equivalent area $\bar{A}$, which is in general a function of the geometry, the stress level and the shape of the load.

5.3. SIMPLIFICATION FOR SPECIAL CASES

Excursus: Problem with the load combination according to ASTM E 1300-04

As stated above, current glass design methods implicitly assume the decompressed surface area to be constant and the major principal stress to be proportional to the load at all points on the surface and during the entire loading history. With these simplifying assumptions, Equation (5.35) can be used to combine loads of different duration. Strangely, Equation (X7.1) of ASTM E 1300-04, which is reproduced in Equation (3.26), is *not* equivalent to Equation (5.35). The author was unable to find an explanation. In his opinion, the equation in the standard is incorrect. This view is supported by a simple example in the following.

There should obviously be no difference between loading a glass element with 10 kN once for 60 s and loading it with the same load twice for 30 s. When calculating the 3 s-equivalent load with Equation (5.35) and $n = 16$, the result is indeed the same for both cases: $q_{3s} = 12.06$ kN. The equation from ASTM E 1300 cited above, however, yields $q_{3s} = 10\,\mathrm{kN} \cdot (60\,\mathrm{s}/3\,\mathrm{s})^{1/16} = 12.06$ kN for the first case and $q_{3s} = 10\,\mathrm{kN} \cdot (30\,\mathrm{s}/3\,\mathrm{s})^{1/16} + 10\,\mathrm{kN} \cdot (30\,\mathrm{s}/3\,\mathrm{s})^{1/16} = 23.10$ kN for the second case.

5.3.3 Common testing procedures

Many common testing procedures use simple geometry and loading. The lifetime prediction model from Chapter 4 can, therefore, be simplified by narrowing its range of validity to constant stress or constant stress rate, uniform stress fields, a constant principal stress ratio and constant crack velocity parameters. Under these conditions, Equation (4.77) can be simplified as follows:

- **Constant, moderate stress rate.** Both coaxial double ring tests (CDR) and four point bending tests (4PB) are usually performed using constant stress rate loading': $\dot{\sigma} = d\sigma/dt = \mathrm{const.}$ Since the principal stress ratio is constant, Equation (5.14) can be used as a starting point for moderate stress rates (considering damage accumulation only, cf. Section 5.2.1). The probability of failure is defined as a function of the major principal stress at failure $\sigma_{1,f}$ (and not of the time) in the following, because this is what is usually measured. Inserting $\sigma_{1,f} = \dot{\sigma}_1 \cdot t_f$ and

$$\sigma_{1,t_0}(t_f) = \left(\frac{\sigma_{1,f}^{n+1}}{\dot{\sigma} \cdot (n+1) \cdot t_0} \right)^{1/n} \tag{5.39}$$

into Equation (5.14) yields (index CSR for constant stress rate):

$$P_f(\sigma_{1,f}) = 1 - \exp\left[-\left(\frac{\sigma_{1,f}}{\theta_{\mathrm{CSR}}} \right)^{\beta_{\mathrm{CSR}}} \right] \tag{5.40}$$

$$\theta_{\mathrm{CSR}} = \left[U(n+1)\dot{\sigma}_f \theta_0^{n-2} \right]^{\frac{1}{n+1}} (A/A_0)^{-1/\beta_{\mathrm{CSR}}} \eta_b^{\frac{-n}{n+1}} \quad ; \quad \beta_{\mathrm{CSR}} = \frac{m_0(n+1)}{n-2} \tag{5.41}$$

The ratio between the scale parameters measured in CDR and 4PB tests respectively is (cf. Equation (5.16)):

$$\frac{\theta_{\mathrm{CSR,CDR}}}{\theta_{\mathrm{CSR,4PB}}} = \left[\frac{2}{\pi} \int_0^{\pi/2} (\cos(\varphi))^{2\tilde{m}} \, d\varphi \right]^{1/\beta_{\mathrm{CSR}}} \tag{5.42}$$

- **Constant stress.** In such tests, the lifetime of specimens exposed to a constant stress is measured. Starting again from Equation (5.14) and inserting

$$\sigma_{1,t_0}(t_f) = \left(\frac{\sigma_1^n \cdot t}{1\,\mathrm{s}} \right)^{1/n} \tag{5.43}$$

yields (index CS for constant stress):

$$P_f(t_f) = 1 - \exp\left[-\left(\frac{t_f}{\theta_{\mathrm{CS}}} \right)^{\beta_{\mathrm{CS}}} \right] \tag{5.44}$$

$$\theta_{\mathrm{CS}} = U \theta_0^{n-2} \sigma_1^{-n} \eta_b^{-n} (A/A_0)^{-1/\beta_{\mathrm{CS}}} \quad ; \quad \beta_{\mathrm{CS}} = \frac{m_0}{n-2} \tag{5.45}$$

- **Inert testing.** The failure probability at inert conditions follows directly from Equation (4.73):

$$P_{f,\text{inert}} = 1 - \exp\left[-\left(\frac{\sigma_{1,f}}{\theta_{\text{inert}}}\right)^{\beta_{\text{inert}}}\right] \quad (5.46)$$

$$\theta_{\text{inert}} = \theta_0 \cdot \eta_{b,\text{inert}}^{-1} \cdot (A/A_0)^{-1/m_0} \quad ; \quad \beta_{\text{inert}} = m_0 \quad (5.47)$$

The ratio between scale parameters measured in inert CDR and inert 4PB tests respectively is (cf. Equation (5.16)):

$$\frac{\theta_{\text{CDR,inert}}}{\theta_{\text{4PB,inert}}} = \left[\frac{2}{\pi}\int_0^{\pi/2}(\cos(\varphi))^{2m_0}\,d\varphi\right]^{1/m_0} \quad (5.48)$$

5.4 Relating current design methods to the lifetime prediction model

The discussion of possible simplifications of the lifetime prediction model in Section 5.3 provides much insight into the issues related to current glass design methods that were identified in Section 3.3. Two issues, however, are still unsolved:

1. European and North American design methods are still not directly comparable.

2. It has not yet been shown how the parameters of these design methods relate to the fundamental parameters of the lifetime prediction model.

Both issues are related since European and North American design methods could easily be compared if their parameters could be related to the fundamental parameters. For the task at hand, it is sufficient to consider an annealed glass element at ambient conditions that is exposed to constant, uniform tensile stress for a relatively long time period. In this very basic case, all assumptions made by current design methods are valid and none of the aforementioned limitations and problematic aspects comes into effect.

European design methods

European design methods are directly based on parameters determined in coaxial double ring tests with constant stress rates (see Section 3.3.2). The fundamental surface condition parameters θ_0 and m_0 can, therefore, be directly calculated from Equations (5.40) and (5.41) with $\eta_b = 1$ if the crack velocity parameters during the tests are known:

$$\theta_0 = \left[\frac{\left(\theta_{\text{CSR}}(A/A_0)^{1/\beta_{\text{CSR}}}\right)^{n+1}}{U \cdot (n+1) \cdot \dot{\sigma}}\right]^{1/(n-2)} \quad ; \quad m_0 = \beta_{\text{CSR}}\frac{n-2}{n+1} \quad (5.49)$$

θ_{CSR} and β_{CSR} are the scale and shape parameters of the Weibull distribution fitted to the constant stress rate test data. In R400 tests (see Section 3.1.2), it is $A = 0.24\,\text{m}^2$ and $\dot{\sigma} = 2 \pm 0.4\,\text{MPa/s}$ (cf. Section 3.1.2).

European design methods use the crack velocity parameters n and S. The parameter n is identical to the one in the lifetime prediction model and v_0 can be determined by $v_0 = S \cdot K_{\text{Ic}}^n$ (from Equations (4.2) and (4.3)).

North American design methods

While details vary between the US and the Canadian standards, they are both based on the glass failure prediction model (GFPM), see Chapter 3. This model is based on two so-called surface flaw

5.4. RELATING CURRENT DESIGN METHODS TO THE LIFETIME PREDICTION MODEL

parameters $\tilde{m}$ and $\tilde{k}$ and one crack velocity parameter n. For a surface area A that is exposed to the uniform equibiaxial t_{ref}-equivalent tensile stress $\sigma_{t_{\text{ref}}}$, Equation (3.21) yields:

$$P_{\text{f}} = 1 - \exp\left\{-\tilde{k} A \sigma_{t_{\text{ref}}}^{\tilde{m}}\right\} \tag{5.50}$$

For the above mentioned conditions, the equivalent reference stress is $\tilde{\sigma} = (A/A_0)^{1/\tilde{m}} \sigma_{t_{\text{ref}}}$ (from Equation (5.22)), such that Equation (5.50) is actually equivalent to Equation (5.23) except for the different reference period. $\tilde{k}$ is analogous to $\bar{k}$ but with $t_{\text{ref}} = 60\,\text{s}$ instead of $t_0 = 1\,\text{s}$, $\tilde{m}$ is identical to the combined parameter $\bar{m}$. The fundamental surface condition parameters θ_0 and m_0 can be calculated, again provided that the crack velocity parameters during the tests are known, using:

$$\theta_0 = \left(\frac{t_{\text{ref}}}{U(\tilde{k}A_0)^{\frac{n-2}{m_0}}}\right)^{1/(n-2)} \quad ; \quad m_0 = \frac{\tilde{m}(n-2)}{n} \tag{5.51}$$

Note: Caution is required concerning the reference time period t_{ref}. As stated in Section 3.2, the reference time period is $t_{\text{ref}} = 60\,\text{s}$ in the GFPM. This t_{ref} was the basis of older US standards and is still used in the current Canadian standard (CAN/CGSB 12.20-M89). Recent versions of ASTM E 1300, however, are based on $t_{\text{ref}} = 3\,\text{s}$ and indicate 3 s equivalent loads in all tables and graphs. Nonetheless, the surface flaw parameters given in ASTM E 1300-04 are still based on a reference time period of 60 s!

5.5 Summary and Conclusions

- **Surface condition and strength data.** According to the lifetime prediction model, the strength of as-received or homogeneously damaged glass specimens should follow a two-parameter Weibull distribution. This is indeed the case with experimental data obtained at *inert* conditions, which reflect the specimen's surface condition only. The goodness-of-fit of *ambient* strength data, however, which are additionally influenced by subcritical crack growth, decreases as time to failure, and thereby the influence of subcritical crack growth, increases. This suggests that the crack velocity parameters are variable, such that they have a barely predictable effect on ambient glass strength. Further investigations on this issue are required and will be presented in Chapter 6.
- **Random surface flaw population.** The random surface flaw population represents the surface condition of as-received glass well. It may, however, often be unrepresentative of in-service conditions, especially if deep surface flaws occur or in the case of machining damage.
- **Risk integral.** Lifetime prediction based on the risk integral implies that the failure probability can be described accurately by accounting for the crack growth (damage accumulation) during the lifetime only while neglecting the influence of the initial surface condition and of the momentary load. Although this approach is the reason for two problems encountered with current models (cf. Section 3.3.4), the risk integral is shown to be a good approximation if the crack depth at failure is substantially greater than the initial crack depth. It is, therefore, suitable for structural design calculations. With very high loading rates or low crack velocities, i. e. when little crack growth occurs, the simplified approach grossly underestimates the probability of failure and leads to unrealistic and unsafe results (resistance above the inert strength). For these cases, which include the interpretation of experiments, the full formulation must be used.
- **Crack growth threshold.** The crack growth threshold of soda lime silica glass can safely be neglected for design as well as for the interpretation of common constant stress rate tests.
- **Biaxial stress fields.** In some special cases (e. g. four point bending tests), the principal stress ratio is time-invariant, such that a correction factor can be used to account for biaxial stress fields. This factor depends on the surface condition parameter m_0, on the crack velocity parameter n and on the principal stress ratio, which is usually location-dependent. For general cases, the following is recommended: For *structural design*, an equibiaxial stress field may be assumed. This assumption greatly simplifies the model and it is safe without being excessively conservative. For the *interpretation of experimental data*, a transient numeric calculation should be performed in order to avoid unsafe results.
- **Simplification of the lifetime prediction model.** Using the above-mentioned findings, it is possible to simplify the lifetime prediction model substantially for structural design (see Section 5.3.1) as well as for common testing procedures (see Section 5.3.3).
- **Avoiding transient analyses.** Analytical considerations show that transient finite element analyses can be avoided even in the case of time-variant loading if (a) the decompressed surface area remains constant and (b) the major principal stress is proportional to the load at all points on the surface (when an equibiaxial stress field is assumed). If both conditions are satisfied, it is possible to apply the load duration effect found for a single crack to a representative stress in a glass element or even to the load. In fact, this is what current design methods do without stating that they do so, although the above-mentioned conditions are not met in all cases that these methods claim to cater for.
- **Current design methods.** It has been shown that current semi-empirical models are actually simplifications and approximations of the lifetime prediction model. The equations derived in this chapter allow model parameters and test data to be converted among the different approaches. Since the parameters of current models combine surface condition and crack growth, this conversion inevitably requires crack velocity parameters to be known or assumed.

Chapter

6

Quantification of the Model Parameters

As is the case with any engineering model, it must be verified whether the lifetime prediction model's input parameters can be determined for in-service conditions with sufficient accuracy and reliability. To this end, a wide range of existing experimental data related to the mechanical resistance of glass are collected, compared and analysed. Furthermore, the crack growth's dependence on the stress rate is assessed using the lifetime prediction model from Chapter 4 and existing large-scale testing data. Issues that need further investigation are identified. They are addressed by specifically designed experiments in Chapter 7.

6.1 Introduction

For several reasons, quantitative studies are crucial in the field of structural glass modelling:

1. The parameters used in contemporary glass design are based on few research projects. This is particularly true in Europe, where a lot of work has been dedicated to the reformulation and simplification of the DELR design method, while its parameters have basically been reused without further discussion (cf. Chapter 3).

2. European design methods were developed independently of North American methods and vice versa. The parameters were not coordinated or compared which is, among other reasons, certainly due to the fact that the conceptual incompatibility of the methods prevents direct comparison of parameters (cf. Section 3.3). The independent parameters of the lifetime prediction model presented in Chapter 4 and summarized in Table 6.1 redress this problem.

3. Results from currently available design methods differ considerably.

Table 6.1: Overview of the lifetime prediction model's input parameters.

Symbol	Designation	Main influence(s)
K_{Ic}	fracture toughness	material 'constant'
K_{th}	crack growth threshold	environmental conditions
Y	geometry factor	geometry of the crack and the element, stress field
v_0, n	crack velocity parameters	environmental conditions, stress rate
θ_0, m_0	surface condition parameters (RSFP)	glass surface condition
a_i	initial depth of a surface crack (SSF)	hazard scenario, glass type
P_f	(target) failure probability	design: target value; with a given action history: result of predictive modelling

6.2 Basic fracture mechanics parameters

Fracture toughness

The fracture toughness K_{Ic} can be considered to be a material constant. It does not depend significantly on influences other than the material itself. If the 10 original data points from Griffith (1920) are converted to stress intensities at failure using Equation (4.6), a relatively uniform value of $K_{Ic} = 0.45\,\text{MPa}\,\text{m}^{0.5}$ is found (Porter, 2001). Modern soda lime silica glasses have a somewhat different chemical composition, which leads to a higher fracture toughness. Table 6.2 gives an overview of published results. In conclusion, a value of $K_{Ic} = 0.75\,\text{MPa}\,\text{m}^{0.5}$ is a good choice for all practical purposes.

Table 6.2: Fracture toughness K_{Ic} of soda lime silica glass at room temperature.

Source	K_{Ic} (MPa m$^{0.5}$)
Wiederhorn (1967)	0.82
Atkins and Mai (1988)	0.78
Gehrke et al. (1990)	0.78
Menčík (1992); from a review of published data	0.72 – 0.82
Ullner (1993)	0.76

Crack growth threshold

Like the fracture toughness, the crack growth threshold K_{th} is known to a fairly high degree of precision, at least in water. Table 6.3 shows a choice of published data from the last 30 years. For in-service conditions, the dependence on the pH value (cf. Section 4.1.5) may be significant. This has, however, no direct consequences because the threshold may be neglected anyway for design (see Sections 4.1.5 and 5.2.2).

Table 6.3: Stress corrosion limit K_{th} of soda lime silica glass at room temperature.

Source	K_{th} (MPa m$^{0.5}$)	Environment
Wiederhorn and Bolz (1970)	0.2	in water, 25 °C
Simmons and Freiman (1981)	0.27	
Gehrke et al. (1990)	0.26	in water, 23 °C
Sglavo and Bertoldi (2004)	0.21	in deionized water

Geometry factor

The geometry factor is in general a function of the stress field, the crack depth, the crack geometry and the element geometry (cf. Section 4.2.1). For the following reasons, however, it makes sense to ignore this dependence:

- The crack growth that affects an element's lifetime occurs at crack depths that are very small in comparison with the element's thickness.

- The depth and geometry of natural flaws are extremely variable. It does not make sense to increase the complexity of the model considerably to achieve a gain in accuracy that would be very small compared to these unavoidable uncertainties.

6.2. BASIC FRACTURE MECHANICS PARAMETERS

Table 6.4:
Overview of experimentally and analytically determined values for the geometry factor of surface flaws on glass.

Type of flaw	Geometry factor Y
Glass on glass scratching*	0.564
Vickers indentation*	0.666
Half-penny shaped crack in a semi-infinite specimen	*0.637–0.663*[†]
Half-penny shaped crack on a flexure specimen	0.713
Quarter-circle crack on glass edges[‡]	0.722
Sandpaper scratching*	0.999
Long, straight-fronted plane edge crack in a semi-infinite specimen	1.120

[*] from Ullner (1993); Ullner and Höhne (1993)
[†] deepest point of the crack; see main text for details
[‡] according to Porter (2001)

Table 6.4 gives some experimentally determined values for the geometry factor (non-italicized text). It remains, however, unknown to what extent these experiments represent actual flaws as they are encountered on glass elements under in-service conditions. Furthermore, it is difficult to separate the geometry factor's influence from other influences in tests, which is why experimental results must be interpreted with care.

To complement the experimental results, the geometry factor shall additionally be estimated using linear elastic fracture mechanics. Because of the extreme brittleness of glass, even elements that are exposed to very small loads fail immediately as soon as a surface crack grows to more than a few tenths of a millimeter. The following conditions of fracture mechanical relevance are, therefore, fulfilled in the case of macroscopic cracks on the surface of glass elements for structural engineering applications (for terminology, see Figure 4.4):

1. The crack depth is small compared to the crack length.
2. The crack depth is small compared to the material thickness.
3. The radius of the crack front (not the crack tip) is substantially larger than the crack depth.
4. The crack depth is negligibly small compared to the overall dimensions of the structural element.

This corresponds to the basic case of a *long, straight-fronted plane edge crack* in a semi-infinite specimen which has a geometry factor of $Y = 1.12$ (Irwin et al., 1967). This value was used by Blank (1993) and subsequently by all European work that is based on this work (cf. Chapter 3).

Other researchers modelled glass surface damage as *half-penny shaped cracks*. Some popular solutions for the geometry factor of such flaws are presented below.

- *Irwin* gives the following analytical geometry factor equation for semi-elliptical surface cracks in an infinite plate (Irwin, 1962)

$$Y(\phi) = \frac{1}{\varphi}\left[\left(\frac{a}{c}\right)^2 \cos^2\phi + \sin^2\phi\right] \qquad \varphi = \int_0^{\pi/2}\left[\left(\frac{a}{c}\right)^2 \cos^2\phi + \sin^2\phi\right]^2 d\phi \quad (6.1)$$

where a is the crack depth, c is half the crack length and ϕ is the parametric angle of the point of interest on the crack front ($\phi = 0$ on the surface, $\phi = \pi/2$ at the deepest point of the crack). For half-penny shaped cracks, a and c are equal, and so $Y = 2/\pi = 0.637$ for all ϕ.

- More complex equations define the geometry factor as a function of the above-mentioned angle ϕ. Tada et al. (1985) gives

$$Y = \frac{2}{\pi}\left(1.211 - 0.186\sqrt{\sin\phi}\right) \qquad (10° < \phi < 170°) \quad (6.2)$$

which yields $Y(10°) = 0.722$ and $Y(\pi/2) = 0.653$. Newman and Raju (1979), finally, predict

$$Y = \frac{1.04}{\sqrt{2.464}}\left[1 + 0.1\left(1 - \sin\phi\right)^2\right] \quad (6.3)$$

which yields $Y(0) = 0.729$ and $Y(\pi/2) = 0.663$.

- Reid (1991), Fischer-Cripps and Collins (1995) and Overend (2002) use the geometry factor of a half-penny shaped crack on a flexure specimen, i.e. $Y = 1.12 \cdot 2/\pi = 0.713$ (e.g. Atkins and Mai (1988); Lawn (1993)).

In order to put them into the context of the experimental values discussed before, the above-mentioned geometry factors are also listed in Table 6.4 (italic text).

Surface flaws on glass edges and at holes are likely to have different geometries from those that flaws on the surface have. Based on theoretical considerations, Porter (2001) proposed modelling flaws on glass edges as quarter circle cracks. He assumed these cracks to maintain their shape during crack growth and used a geometry factor of $Y = 0.722$.

From a practical point of view, half-penny shaped cracks are not very likely to occur on glass surfaces. Particularly flaws caused by hard contact are likely to be long, such that their geometry factor can be expected to lie close to $Y = 1.12$. This assumption is supported by the value of 0.999 that was determined for sandpaper scratching (Ullner, 1993; Ullner and Höhne, 1993). A geometry factor of $Y = 1.12$ is, therefore, a sensible assumption for surface cracks away from edges. On glass edges containing machining flaws (i.e. flaws that are introduced by any kind of machining such as edge working or drilling), half-penny or quarter-circle cracks may be more appropriate choices.

This is an area about which more insight and more reliable information are required. To obtain them, it would be very useful if the geometry factor of close-to-reality surface flaws could be determined experimentally. This is investigated in Chapter 7.

6.3 Crack velocity parameters

Available Data

Table 6.5 shows the 'classic' European crack velocity parameters on which all European work on the design of structural glass elements is based (cf. Section 3.3.4). They were published in Kerkhof et al. (1981) and used the ambient condition crack velocity parameters from Richter (1974). He determined those parameters by optically measuring the growth of large through-thickness cracks on the edges of specimens loaded in uniform tension (see Section 3.1.3).

Environment	n (-)	v_0 (mm/s)	S (m/s)·(MPa m$^{0.5}$)$^{-n}$
water*	16.0	50.1	5.0
air, 50% relative humidity*	18.1	2.47	0.45

Table 6.5: Crack velocity parameters for soda lime silica glass according to Kerkhof et al. (1981) (converted), based on experimental data from Richter (1974).

Based on these values, the design parameters given in Table 6.6 were chosen in Blank (1993) and used for the DELR design method. Table 6.7 summarizes values that were determined in dynamic four point bending tests by nine different laboratories on a large sample of glasses from Saint-Gobain (Ritter et al., 1985). Finally, Table 6.8 gives an overview of a selection of more recent experimental data. Some sources give values for n only, these are listed in Table 6.9. The v_0 values given in the tables are calculated from S assuming $K_{Ic} = 0.75$.

Environment	n (-)	v_0 (mm/s)	S (m/s)·(MPa m$^{0.5}$)$^{-n}$
Laboratory testing and 'Summer conditions'	16.0	4.51	0.45
'Winter conditions, melting snow, 2°C'	16.0	8.22	0.82

Table 6.6: Crack velocity parameters used in the DELR design method.

Figure 6.10 gives a complete overview of all data. It contains the v-K_I-curves resulting from all v_0 and n values in the above mentioned tables, some published data sets and the v-K_I-curve proposed by

6.3. CRACK VELOCITY PARAMETERS

Table 6.7:
Crack growth data for SLS glass in water, mean values from 9 laboratories and 2000 specimens (70x25x1 mm) (Ritter et al., 1985).

Failure mode	n (–)	v_0^* (mm/s)
all	16.8	1.78
edge failure	15.5	0.548
surface failure	17.70	10.7

* calculated with $Y = 1$

Table 6.8: Recent crack growth data from Vickers indentation experiments on SLS glass.

Environment	n (–)	v_0 (mm/s)		Test setup (Source)
Water	26 ±7	$3.7 \cdot 10^7$		dynamic fatigue tests (Sglavo and Bertoldi, 2004)
Water	18 ± 1	19 ± 4		cyclic fatigue tests, as-indented and annealed (Sglavo et al., 1997)
Water	20.1 ± 0.7	28.8 ± 6.4	*	dynamic fatigue tests
	19.9 ± 0.7	6.4 ± 1.4	†	(Sglavo and Green, 1999)
'normal' environment (air, 27 °C, 65% RH)	19.7 – 21.2	0.2 – 0.4		direct optical measurement
	21.8	2.6		4PB - dynamic fatigue, natural flaws
	21.1	2.4		4PB - dynamic fatigue, indentation flaws (Dwivedi and Green, 1995)

* as-indented   † annealed

Table 6.9:
Additional data for the crack velocity parameter n.

Source	Glass type	n (–)
Schneider (2001) from dynamic fatigue tests	SLS	17 – 21
Fink (2000), static fatigue tests in the open air and in a climate chamber	SLS	16
Choi and Holland (1994)*, dynamic fatigue tests	SLS	16.4
Menčík (1992), from a review of published data	SLS	12 – 17
Menčík (1992), from a review of published data	BSG	24 – 35

* data reproduced in Choi et al. (1997)

Ullner (1993)[1]. When modelling subcritical crack growth, the v-K_I-relationship is generally assumed to be valid over the full K_I-range (cf. Sections 4.1.3 and 4.2.4 and 5.2.2). This is why the curves that represent design models extend to the entire range of the figures' axes.

How reliable is this data?

The large scatter of the data is in itself a source of doubt about the reliability of crack velocity parameters. Furthermore, the following issues should be considered:

- Crack growth data determined from measurements on large through-thickness cracks may not accurately represent the behaviour of natural flaws. Recent research shows that the growth of large cracks depends strongly on the quality of annealing (Gy, 2003). Residual stresses, however small, have a strong influence on the stress intensity factor of large cracks and can cause an overestimation of the crack velocity by a factor of ten (Figure 6.11).

- Indentation flaw based methods can be expected to represent short-crack behaviour more accurately. But even then, residual stress fields can strongly influence susceptibility to stress

---

[1] This curve is not straight because the model is based on the exponential function given in Equation (4.4) instead of the power law in Equation (4.2).

84 CHAPTER 6. QUANTIFICATION OF THE MODEL PARAMETERS

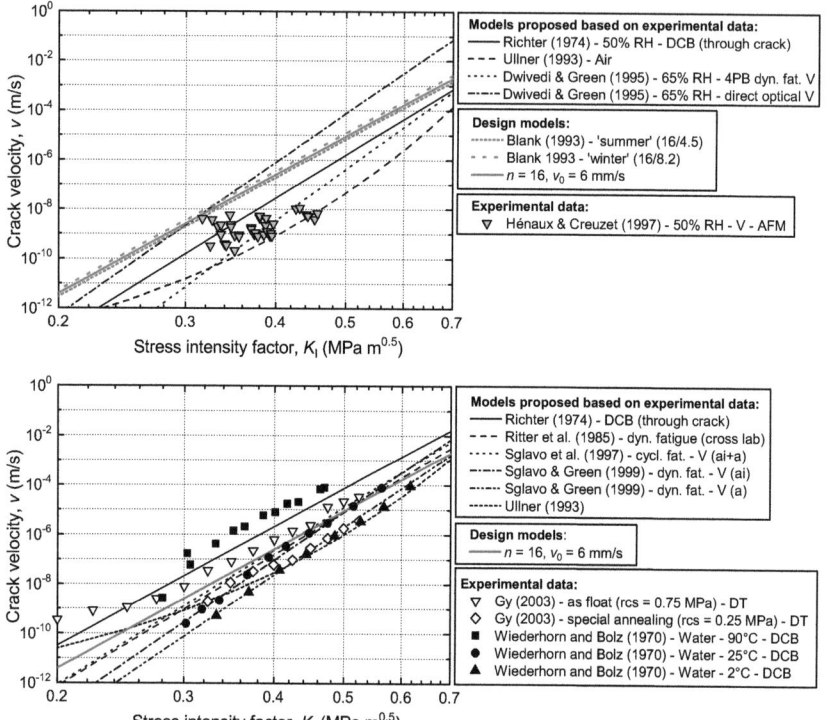

Figure 6.10: Crack growth data overview in air (above) and in water (below). Abbreviations: V = Vickers indentation, ai = as indented, a = annealed, DT = double torsion (test type), DCB = double cantilever beam (test type), dyn. fat. = dynamic fatigue, rcs = residual core stress.

corrosion. It is unclear whether as-indented or annealed specimens represent the natural flaw state more closely.

♦ It would be safer but also far more time-consuming to determine crack velocity parameters on specimens in their in-service conditions and with load durations similar to typical service lifetimes.

A seemingly interesting option in view of the difficulties in defining reliable crack velocity parameters is to assume a conservative value. For design, this approach is safe and unproblematic. However, when deriving surface condition parameters from strength data obtained at ambient conditions, it is inadvisable. Overestimating crack growth during tests means underestimating the initial flaw depth. The resulting surface flaw parameters are too optimistic and thus unsafe.

Conclusions

⇨ The scatter of available crack growth data is extremely large.

⇨ More data and larger samples are available for glass immersed in water than for glass in air. Immersion in water can be considered as a worst case scenario for structural engineering applications.

⇨ Measurements in water fit well to a linear relationship on a log-log scale and thus confirm the crack growth law from Equation (4.3).

6.3. CRACK VELOCITY PARAMETERS

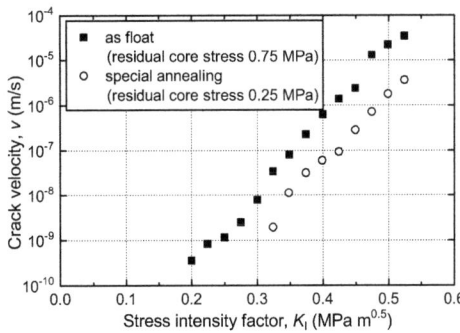

Figure 6.11:
Influence of the quality of annealing on the v-K_I-curve obtained with double torsion samples (data from Gy (2003)).

▷ Although the parameters in Table 6.6 were chosen in Blank (1993) based on those in Table 6.5, they represent substantially higher crack velocities.

▷ The available data indicate, at least as a tendency, that the main difference between the measurements is a parallel shift of the v-K_I-curve. This is consistent with the finding that the main consequence of differences in humidity and temperature is such a parallel shift (see Section 4.1). At a given temperature, the scatter is larger and the crack velocity slower in air than in water. This suggests that the diffusion rate of humidity might influence the crack velocity in air. If a shortage in the supply of water slows down crack growth, the crack velocity parameters v_0 and n might be stress rate dependent. This is an important issue that needs to be further investigated, see Section 6.5.

▷ *For design,* a constant value of $n = 16$ is a reasonable assumption for all environmental conditions. For general applications, $v_0 = 6$ mm/s may be used as a conservative assumption (cf. Figure 6.10). For glass elements that are permanently immersed in water, a higher value of e.g. $v_0 = 30$ mm/s might be required. Further differentiation of environmental conditions, e.g. considering summer and winter conditions, is not recommended for modelling purposes. The potential difference between the two cases is very small compared to the scatter of the data. The definition of two parameter sets would therefore be rather arbitrary and would not necessarily increase the accuracy of the model. The complexity of the calculation process, on the other hand, would be increased considerably.

▷ *For the interpretation of test data,* one must proceed with caution. Strength data from tests at ambient conditions are inevitably dependent on the surface condition *and* on crack growth behaviour. In view of the above conclusions, this is a major drawback:

- The large scatter of the crack velocity parameters makes it very difficult to obtain accurate surface condition information from tests at ambient conditions.
- Inaccurate estimation of the crack velocity during testing can yield unsafe design parameters.

Laboratory testing at *inert* conditions would, therefore, be preferable.

6.4 Ambient strength data

If the equations from Section 5.4 are used, the surface condition parameters θ_0 and m_0 can be derived from existing design methods and published laboratory test data. This enables a consistent comparison despite the different nature of the design methods and the data. As mentioned before, the parameters that govern crack growth during the tests must be estimated to interpret data obtained at ambient conditions. Based on the findings in Sections 6.2 and 6.3, $n = 16$, a series of different v_0 values, $Y = 1.12$ and $K_{Ic} = 0.75 \text{ MPa m}^{0.5}$ were used to calculate the surface condition parameters shown in Table 6.12. It can be seen that the crack velocity parameter v_0 has a significant influence on the result. Since existing test data do not allow the most appropriate v_0 value to be determined with confidence (cf. Section 6.3), further investigation was definitely required and is done in Section 6.5.

The data in Table 6.12 were determined using specimens with very different surface areas A exposed to tensile stress. This means that in order to obtain θ_0, test data had to be converted from the surface area used in the tests to the reference surface area of $A_0 = 1.0 \text{ m}^2$ to account for the size effect. The fact that similar θ_0 and m_0 values were obtained confirms the size effect for as-received glass and glass with artificially induced homogeneous surface damage.[2] Unsurprisingly, the scatter of the weathered glass surface condition parameters is large (cf. Section 5.1).

Figure 6.13 gives a visual comparison of the surface condition data obtained for $v_0 = 0.01 \text{ mm/s}$.[3]

Table 6.12: Surface condition parameters determined from laboratory tests at ambient conditions unless otherwise stated. Crack velocity parameters: $n = 16$, v_0 as given in the table header. Note: m_0 does not depend on v_0.

	A (cm²)	θ_0 assuming $v_0 =$				m_0 (-)
		0.01 mm/s	0.1 mm/s	1 mm/s	6 mm/s	
		(MPa)				
As-received glass						
DIN 1249-10:1990	2400	62.89	74.13	87.38	99.31	4.94
Brown (1974)		70.21	82.76	97.55	110.87	6.39
Beason and Morgan (1984)		60.27	71.04	83.74	95.17	7.88
Fink (2000)	23.8	61.20	72.15	85.04	96.66	6.30
Ritter et al. (1985)*	6.25	36.70	36.70	36.70	36.70	6.14
This study from ORF data*†		62.95	62.95	62.95	62.95	8.09
This study from inert tests*‡	20.4	67.57	67.57	67.57	67.57	7.19
Weathered window glass						
Beason (1980)		27.65	32.59	38.42	43.67	5.25
ASTM E 1300-04		40.93	48.25	56.87	64.64	6.13
Fink (2000)	23.8	20.82	24.55	28.93	32.88	3.76
Glass with artificially induced homogeneous surface damage						
DELR design method / prEN 13474	2400	28.30	33.36	39.32	44.69	20.59
Blank (1993)	2400	35.37	41.70	49.15	55.86	33.19
Blank (1993)	2.54	33.29	39.24	46.26	52.57	23.53

* Inert conditions, therefore independent of the crack velocity parameters.
† See Section 6.5.
‡ See Section 7.3.

---

[2] The Ritter et al. (1984) data do not seem to fit. There is, however, a straightforward explanation for this. The specimens are so tiny that they would, according to the statistical size effect, contain only extremely shallow flaws. In reality, even tiny specimens undergo some handling damage. They do, therefore, not reach the extremely high resistance suggested by statistics.

[3] The choice of this v_0 value seems arbitrary at the moment. It is explained in Section 6.5 based on the findings of that section.

6.4. AMBIENT STRENGTH DATA

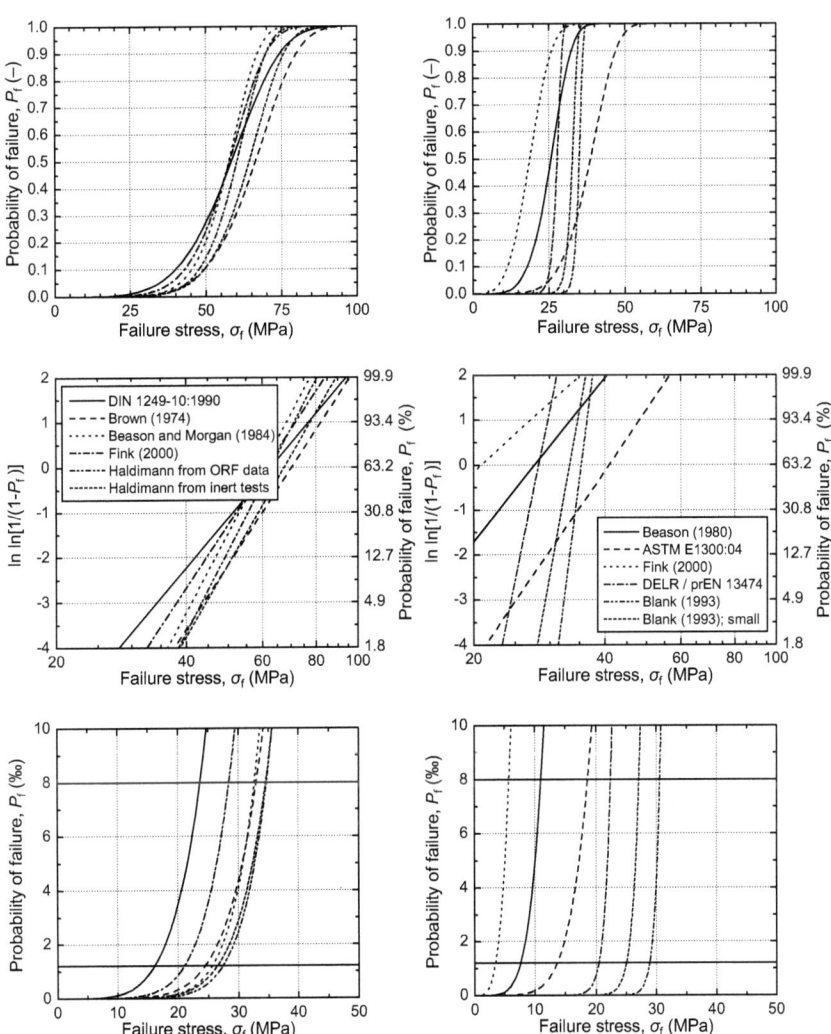

Figure 6.13: Comparison of surface condition parameters for as-received glass (left) and for weathered window glass and glass specimens with artificially induced homogeneous surface damage (right). Top: cumulative distribution function (CDF), middle: Weibull plot, bottom: CDF for $0 \leq P_f \leq 10‰$. The horizontal lines in the figures at the bottom represent the P_f values used in European (1.2‰) and North American (8‰) standards.

6.5 Analysing large-scale experiments

Introduction and aim

The preceding sections have shown that:

1. The conditions under which crack growth and surface condition data are determined should be as close as possible to in-service conditions.
2. The crack velocity parameters vary widely and are possibly load rate dependent.
3. While surface condition measurements at ambient conditions are inevitably influenced by subcritical crack growth, this is not the case at inert conditions. Surface condition data would, therefore, be more reliable if they were obtained at inert conditions.

These issues are investigated in the present section. More specifically, the objectives are to:

- estimate crack growth's dependence on the stress rate at ambient conditions;
- determine surface condition parameters from a test setup that is at the same time representative of in-service conditions and independent of the unknown crack growth behaviour during the tests.

The lifetime prediction model developed in Chapter 4 enables any test setup to be consistently modelled. No particularly simple geometry or stress field is required and the model is valid over the full range of possible parameters, including very slow crack growth or no crack growth at all. The above-mentioned objectives can therefore be achieved using this model and published experimental data that are particularly well suited for the task at hand.

The data used come from a large test programme conducted in Canada by the Ontario Research Foundation (ORF) (Johar, 1981, 1982). Rectangular annealed float glass specimens with a thickness of 6 mm and a panel size of 1.525 m × 2.440 m (3.721 m^2) were tested at ambient conditions. A uniform lateral load was applied to the panels. The displacement perpendicular to the plates was restrained along the plate edges, while rotation remained unconstrained. The data were chosen for two reasons:

- The specimen dimensions are comparable to those commonly used in architectural applications.
- Tests were performed at load rates ranging from 0.0025 kPa/s to 25 kPa/s. This results in times to failure ranging from less than a second to half an hour, which is ideal for investigating the crack growth speed's dependence on the load rate.

The laboratory testing was performed in two phases. In Phase II, glass panes from three different manufacturers were tested at load rates of 0.15 kPa/s, 1.5 kPa/s and 15 kPa/s. In Phase III, glass panes from one manufacturer were tested at load rates of 0.0025 kPa/s, 0.025 kPa/s, 0.250 kPa/s, 2.5 kPa/s and 25 kPa/s. Figure 6.14 shows an overview of the experimental data that are used in the following[4]. Further details and tables are provided in Appendix D.

Edge failures are not representative of the task at hand. Therefore, only specimens with failure origin on the glass surface were used. Two series, those at 2.5 kPa/s and at 25 kPa/s, are not shown in the figure. Their fit to any smooth curve is particularly poor. While this is not surprising in view of the relatively small sample size and the high scatter of the effective pressure rate, it makes any results derived from the data very sensitive to the choice of the fitting algorithm.

Concept

Test results at ambient conditions reflect a combination of all resistance parameters (see Table 6.1). In light of information presented in preceding chapters, the fracture toughness and the geometry factor are taken to be $K_{Ic} = 0.75$ MPa m$^{0.5}$ and $Y = 1.12$. The parameters v_0, n, θ_0 and m_0 remain unknown. As fitting to experimental data cannot provide all these parameters, some additional a

---
[4] The choice of the estimator for the empirical probability that is used to plot the experimental data is discussed in Section E.2.

6.5. ANALYSING LARGE-SCALE EXPERIMENTS

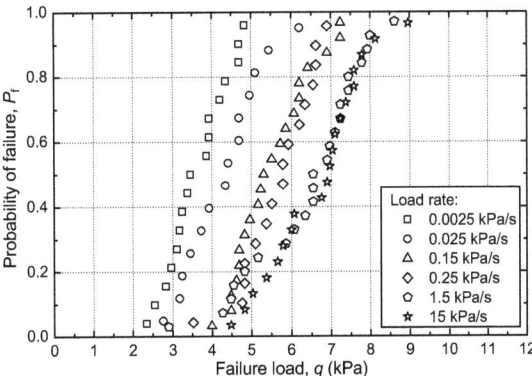

Figure 6.14:
Ontario Research Foundation data, overview (data sources: Johar (1981, 1982)).

priori knowledge is required. It can be seen in Figure 6.14 that there is no appreciable increase in the measured strength between the load rates of 1.5 kPa/s and 15 kPa/s. In other words, the strength is load rate-independent at these high load rates. This implies that no subcritical crack growth occurs and, consequently, that the test data represent the inert strength. To further verify this, Figure 6.15 additionally shows the (otherwise unused, cf. above) data sets with load rates of 2.5 kPa/s and 25 kPa/s. Despite the large scatter of the data, it can be seen clearly that even at 25 kPa/s, strength does not increase significantly compared to 1.5 kPa/s. In conclusion, the data set at 15 kPa/s may be assumed to represent the inert strength of as-received glass specimens. This assumption is of major importance for the calculations that follow. It therefore requires further validation. To this end, experiments, in which subcritical crack growth is prevented, were performed within the present work (see Section 7.3). The results confirm the aforementioned assumption.

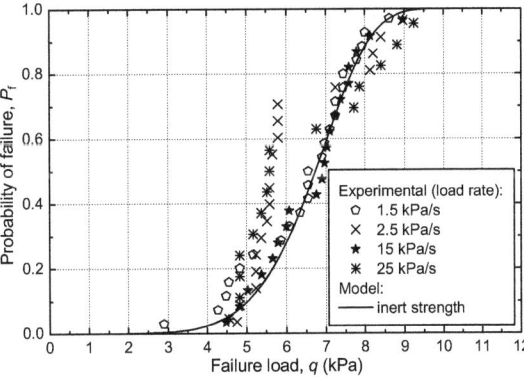

Figure 6.15:
Does ambient strength data converge on the inert strength at very high load rates?

The inert strength is independent of the crack velocity parameters (v_0, n). Therefore, the surface condition parameters θ_0 and m_0 can readily be found by fitting the lifetime prediction model to the test data.

In conclusion, the combination of the Ontario Research Foundation experimental data with the lifetime prediction model allows the following tasks to be carried out:

- In a first phase, the surface condition parameters of large as-received glass specimens can be determined in quasi-inert conditions, that is without the unwanted influence of subcritical crack growth.
- Assuming that the surface condition of all specimens is similar, this information can in a second phase be used to determine the crack growth speed as a function of the load rate.

Procedure

The modelling procedure and data flow are very similar to the design process explained in Section 8.2. The only difference is that for the specific task at hand, input parameters of the material model are determined by fitting model output to experimental failure probability data.

Structural analysis was done using the commercial finite element software package Abaqus (ABAQUS 2004). The key features of the FE model are as follows: rectangular glass plate of dimensions 1525 mm × 2440 mm × 5.8 mm, translation restrained along all four edges; 520 finite elements (required because the model is very sensitive to stress concentrations); uniform lateral load, increasing linearly in 200 steps from 0 kPa to 40 kPa; non-linear calculation; 8-node shell elements with quadratic interpolation (S8R). Pre-processing and post-processing scripts were developed by the author to generate the model and to extract the input data for the lifetime prediction model from the FE model output. (Major and minor in-plane principal stresses for all faces of all finite elements at all load steps and the surface area of all finite elements are required.)

For the large model at hand, the failure probability calculation is computing time intensive. It is, therefore, of vital importance to use an efficient fitting algorithm. At the time of writing, there are no mature open source minimizing algorithms in C# available. In order to avoid the time-consuming implementation of such an algorithm, the lifetime prediction model was implemented in Matlab (Matlab 2005), a commercial technical computing software package. This allows Matlab's advanced built-in algorithms to be used. The Matlab implementation contains the following main components:

- a simple action history generator for linearly increasing actions,
- an implementation of the lifetime prediction model (see Section 8.5.2),
- fitting algorithms that allow the fitting of the model to experimental data using the maximum likelihood method and the least squares method based on ΔP_f or Δaction,
- various helper functions (experimental data input, empirical probability of failure, read output data of finite element analyses, etc.),
- scripts to automate the input/output and calculation process.

The source code of some key functions is provided for reference in Section G.1. The fitting methods are explained in Section E.3. For general aspects of the software implementation and a performance comparison between C# and Matlab, see Section 8.5.

Results

With maximum log-likelihood fitting[5], the following surface condition parameters were found from the data set at 15 kPa/s (curve in Figure 6.15):

$$\theta_0 = 63 \text{ MPa} \quad ; \quad m_0 = 8.1$$

Assuming that these surface condition parameters are similar for all specimens, the variance in the crack growth speed as a function of the stress rate can now be assessed. As an infinite number of (v_0, n) combinations would provide an equally good fit to failure load data, some a priori knowledge has to be introduced. This is possible on the basis of Section 4.1 and Section 6.3, in which the main variability of the v-K_I-relationship has been found to be a parallel shift of the curve. A fixed value of $n = 16$ (representing the slope of the v-K_I-relationship) was therefore used. This enabled the linear crack velocity parameter v_0 to be determined for each load rate by fitting the lifetime prediction model to experimental data. Figure 6.16 shows the experimental data together with the least squares fits.

Figure 6.17 shows the v_0 values that were obtained with the three fitting algorithms (see Section E.3). In order to increase the expressiveness of the graph, the x-axis shows the stress rate ($\dot{\sigma}$) in the tests rather than the load rate as in previous figures. Since the load/stress relationship is

---

[5] Least-squares fitting based on ΔP_f or Δaction yields similar results.

6.5. ANALYSING LARGE-SCALE EXPERIMENTS

not perfectly linear, a *mean* stress rate is indicated. The results depend considerably on the fitting algorithm. Nevertheless, it can be seen clearly that the crack velocity parameter v_0 is strongly stress rate dependent. On double logarithmic scales, the v_0-$\dot{\sigma}$-relationship is approximately linear. At the stress rate of $\dot{\sigma} = 2\,\text{MPa/s}$, which is commonly used in laboratory tests, v_0 is about $10\,\mu\text{m/s}$. This result is confirmed by Table 6.12: Assuming $v_0 = 10\,\mu\text{m/s}$ produces the best agreement between surface condition parameters θ_0 that are derived from tests at ambient and at inert conditions respectively.

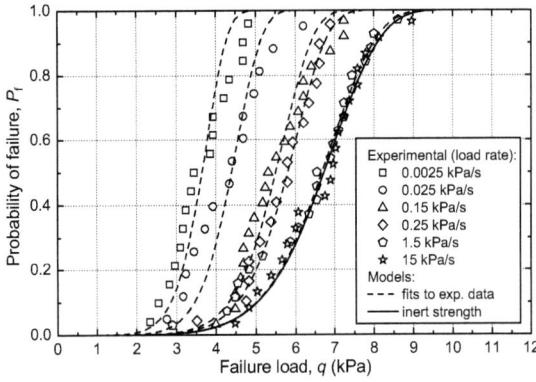

Figure 6.16:
Comparison between experimental data and fitted models.

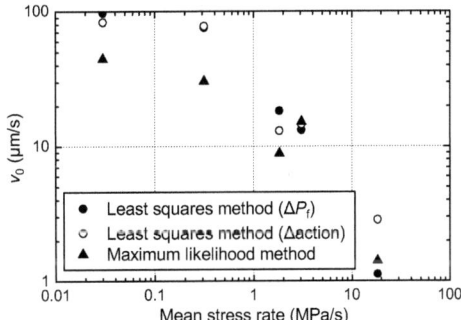

Figure 6.17:
Crack velocity parameter v_0 obtained when fitting the lifetime prediction model to the Ontario Research Foundation experimental data.

Conclusions

⇨ The lifetime prediction model from Chapter 4 was applied successfully in order to analyse existing experimental data from tests with a complex time-dependent stress field at ambient conditions.

⇨ By using the lifetime prediction model and data from tests at very high load rates, it was possible to obtain surface condition parameters from a test setup that is at the same time representative of in-service conditions and independent of unknown crack growth behaviour. The parameters found are $\theta_0 = 63\,\text{MPa}$ and $m_0 = 8.1$. They should, however, be further verified by laboratory testing. This is performed and discussed in Chapter 7.

⇨ By using the above-mentioned surface condition parameters, the crack velocity parameters in tests at moderate stress rates were estimated. The results support the hypothesis presented in Section 6.3, i.e. that the v-K_I-relationship in laboratory tests is strongly stress rate dependent. At the stress rate of $\dot{\sigma} = 2\,\text{MPa/s}$, which is commonly used in laboratory tests, v_0 is found to be about $10\,\mu\text{m/s}$.

6.6 Residual surface stress data

The residual surface stress σ_r is a particularly interesting parameter because it is:

- independent of the surface condition as well as of loading and environmental conditions.
- much higher than the glass's inherent strength.

This means that the potential gain in design strength is very important if a high level of σ_r can be guaranteed, e. g. by quality control measures. In order to quantify the order of magnitude of this potential gain and to estimate the residual stress that may be assumed for design without quality control measures, available residual stress data are analysed in the following. Published residual stress data from Luible (2004b), Schneider (2001) and Laufs (2000a) as well as measurements by the author (see Chapter 7) are used. All data represent residual stress away from edges.

Statistical information is provided in Tables 6.18 and 6.19 and a graphical overview of the data is in Figures 6.20, 6.21 and 6.22. The fit of the data to a normal distribution is rather poor, but the fit to a log-normal distribution is not significantly better. With statistical techniques such as tail-approximation, a better fit at the lower tail can be achieved, but at the expense of a worse fit near the mean. This in unproblematic for design, but undesirable for modelling purposes. The normal distribution is conservative at the lower tail (see Figure 6.22).

Thickness	(mm)	6	8	10	ALL HSG
Specimens	(–)	20	61	75	156
Min	(MPa)	64.3	52.4	42.5	42.5
Max	(MPa)	86.7	85.1	66.5	86.7
Mean	(MPa)	72.1	62.5	49.3	57.4
Standard deviation	(MPa)	6.12	6.13	5.36	10.15
Coefficient of variation	(%)	8.49	9.81	10.89	17.69
5% fractile (normal dist.)	(MPa)	62.1	52.4	40.4	40.7
2% fractile (normal dist.)	(MPa)	59.6	49.9	38.2	36.5

Table 6.18: Statistical analysis of heat strengthened glass residual stress data (data sources: Laufs (2000a); Luible (2004b); Schneider (2001) and measurements by the author).

Thickness	(mm)	6	8	10	15	ALL FTG
Specimens	(–)	20	9	503	65	597
Min	(MPa)	109.2	98.2	93.2	93.2	93.2
Max	(MPa)	128.5	114.5	192.1	173.6	192.1
Mean	(MPa)	121.8	105.1	137.0	121.3	134.3
Std deviation	(MPa)	5.09	5.75	23.34	17.08	23.11
Coeff. of variation	(%)	4.18	5.47	17.03	14.08	17.21
5% fractile (N)	(MPa)	113.4	95.6	98.6	93.2	96.3
2% fractile (N)	(MPa)	111.3	93.2	89.1	86.2	86.8

Table 6.19: Statistical analysis of fully tempered glass residual stress data (data sources: Laufs (2000a); Luible (2004b); Schneider (2001) and measurements by the author).

The residual stress data allows the following conclusions to be drawn:

⇨ The residual surface stress in heat treated glass varies widely among specimens as well as among manufacturers[6]. This means that a substantial gain in design strength can be obtained by ensuring a high residual stress level, e. g. by quality control measures.

⇨ The residual surface stress of heat strengthened glass seems to be inversely proportional to the glass's thickness.

⇨ When defining a characteristic value of the residual stress as the 5%-fractile value[7], $\sigma_{r,k,HSG}$ = 40 MPa and $\sigma_{r,k,FTG}$ = 95 MPa are obtained (rounded values from Table 6.18 and Table 6.19). These values are substantially higher than those in prEN 13474-1:1999 (25 MPa for HSG, 75 MPa for FTG).

---

[6] This is not directly visible in the figures, but is in the original data. It is also the reason why the fully tempered glass data look like a superposition of at least two samples with very different mean values.

[7] This is how virtually all characteristic material properties used in engineering are defined.

6.6. RESIDUAL SURFACE STRESS DATA

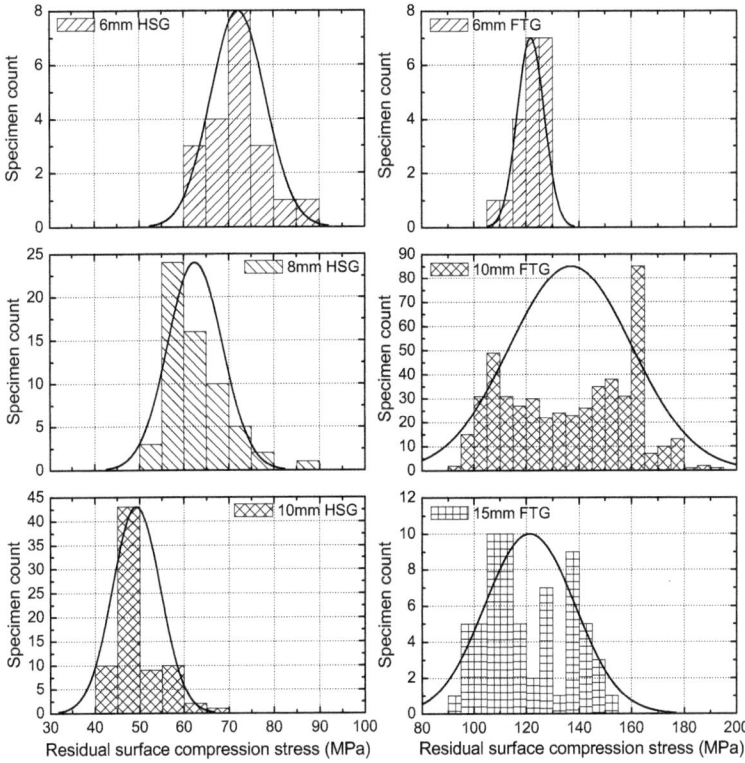

Figure 6.20. Residual stress data, pooled by glass thickness (histograms and fitted normal distributions; data sources: Laufs (2000a); Luible (2004b); Schneider (2001) and measurements by the author).

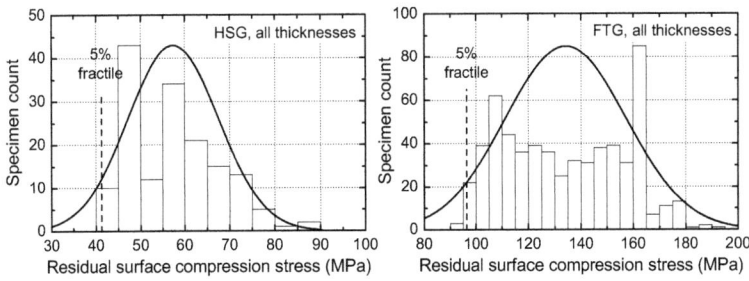

Figure 6.21: Residual stress data, all thicknesses pooled (data sources: Laufs (2000a); Luible (2004b); Schneider (2001) and measurements by the author).

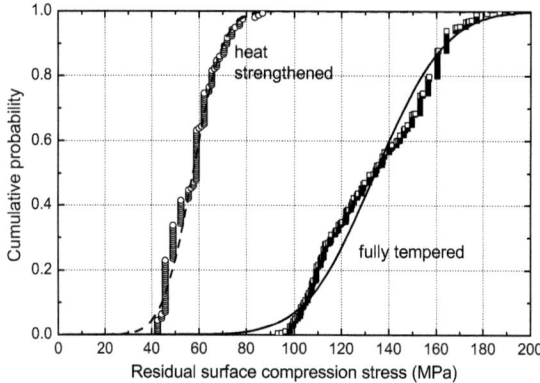

Figure 6.22:
Residual stress data and fitted normal distributions; cumulative.

Excursus: Residual stress on glass edges

It is much more difficult to measure the residual stress on or near glass edges than it is to measure such stress on other parts of the surface. Laufs (2000b) found the residual stresses on edges of fully tempered glass to be about 15%–25% lower than those away from the edges. In heat strengthened glass, on the other hand, the residual edge stress is found to be almost 50% higher than the residual surface stress. No distinct correlation between residual stress on edges and away from edges could be found (Bucak and Ludwig, 2000). This is not surprising, as the residual stress level depends on the temperature distribution in the glass element during the tempering process, which is in turn a function of the element's geometry as well as of the cooling equipment and process.

6.7 Summary and Conclusions

⇨ **Fracture toughness.** The fracture toughness K_{Ic} does not depend significantly on influences other than the material itself and can thus be considered to be a material constant. A value of $K_{Ic} = 0.75\,\text{MPa}\,\text{m}^{0.5}$ is a good choice for all practical purposes.

⇨ **Crack growth threshold.** While the crack growth threshold K_{th} is known to a fairly high degree of precision in water, its dependence on the pH value may be significant for in-service conditions. This has, however, no direct consequences because the threshold may usually be neglected (cf. Section 5.2.2).

⇨ **Geometry factor.** Flaws away from edges are likely to be long compared to their depth and a geometry factor of $Y = 1.12$ is a reasonable assumption. On glass edges containing machining flaws (i. e. flaws that are introduced by any kind of machining such as edge working or drilling), half-penny or quarter-circle cracks with lower geometry factors may be more appropriate choices. Clearly, the geometry factor of natural flaws is an area about which more information is required. To this end, it would be very useful if the geometry factor of close-to-reality surface flaws could be determined experimentally. This is investigated in Chapter 7.

⇨ **Crack velocity parameters.** It has been found that the scatter of the crack velocity parameters is extremely large. In addition to their dependence on environmental conditions, they have been found to be strongly stress rate dependent. For design purposes, a constant exponential crack velocity parameter $n = 16$ and a conservative estimate of the linear crack velocity parameter $v_0 = 6\,\text{mm/s}$ are sensible choices. Testing is discussed below.

⇨ **Obtaining reliable information from experiments.** Strength data from tests at ambient conditions are inevitably dependent on the surface condition *and* on crack growth behaviour. In view of the above conclusion, this is a major drawback. The large scatter and the stress rate dependence of the crack velocity parameters make accurate estimation of the crack growth that occurs during experiments at ambient conditions difficult. Inaccurate estimation, however, can yield unsafe results. Laboratory testing at inert conditions, in which results are not influenced by crack growth, would therefore be preferable.

⇨ **Size effect.** Experimental data confirm the existence of a size effect for as-received glass and glass with a homogeneously scratched surface.

⇨ **Surface condition parameters of as-received glass.** By fitting the lifetime prediction model to existing experimental data from very high stress rate tests on large specimens at ambient conditions, the following surface condition parameters for as-received glass have been found: $\theta_0 = 63\,\text{MPa}$, $m_0 = 8.1$.

⇨ **Research needs.** Experimental investigations are required to answer the following questions:
  - Is inert testing feasible for structural applications? What testing procedure is suitable?
  - Can the above-mentioned surface condition parameters, which were determined from very high stress rate ambient tests on large specimens, be confirmed in truly inert tests on small specimens?
  - How can deep, close-to-reality surface flaws be created for strength testing? What is their mechanical behaviour?

These experimental investigations were conducted as part of the present study and are presented in Chapter 7.

⇨ **Residual stress.** The residual surface stress in heat treated glass varies widely among specimens as well as among manufacturers. This means that a substantial gain in design strength can be obtained by ensuring a high residual stress level, for example by quality control measures.

Chapter

7

Experimental Investigations

This chapter describes experimental investigations that were conducted in order to answer the questions that arose from Chapter 6, namely (a) Inert testing: Is inert testing feasible for structural applications? What testing procedure is suitable? Can the surface condition parameters of as-received glass¦, which were found in Section 6.5, be confirmed? (b) Deep close-to-reality surface flaws: How can deep, close-to-reality surface flaws be created for strength testing? What is their mechanical behaviour? (c) Detectability: What is the probability of detecting surface flaws through today's visual inspections?

7.1 Objectives

The main focus of the present work lies on theoretical considerations of glass strength. Therefore, the aim of the laboratory testing described in this chapter is not to give statistically significant material data (such testing would require more resources in terms of equipment, specimens and time) but to answer the following questions:

- *Inert testing:* Is inert testing feasible for structural applications? What testing procedure is suitable? Can the surface condition parameters of as-received glass that were found in Section 6.5 be confirmed?

- *Deep close-to-reality surface flaws:* How can deep, close-to-reality surface flaws be created for strength testing? What is their mechanical behaviour?

- *Detectability:* What is the probability of detecting surface flaws through today's current visual inspections?

First, a suitable coaxial double ring test setup was developed and its behaviour was verified (Section 7.2). Annealed glass specimens were then tested at various stress rates with and without a hermetic surface coating to assess whether it is possible to achieve near-inert conditions in this manner (Section 7.3). These tests serve furthermore to determine the surface condition parameters of as-received glass for comparison with the values found in Section 6.5. To investigate the mechanical behaviour of deep surface flaws, a surface scratching device was developed and destructive tests were conducted on scratched annealed, heat strengthened and fully tempered glass specimens (Section 7.4). Finally, the probability of detecting surface flaws when looking at the glass from a distance of 3 m was estimated (Section 7.5).

7.2 Test setup and measurements

Choice of a suitable test setup

In current European standards, there are two coaxial double ring test setups. For the task at hand, they are both unsuited for the following reasons:

- The R45 test setup (EN 1288-5:2000) is too small (see Table 7.1). The tolerances and the limited precision of common structural testing equipment would have an important influence at that scale. Furthermore, the conversion of test results to the much larger unit surface of $1\,m^2$ (the basis for the surface flaw parameters) is influenced predominantly by the scatter of the data and would therefore be of limited accuracy and reliability.

- The specimens of the R400 test setup (EN 1288-2:2000) are far too big (see Table 7.1) to fit in a microscope for the flaw depth measurement.

To overcome these problems, a more suitable test setup is required. The choice of a loading ring diameter of 51 mm and a reaction ring diameter of 127 mm offers an ideal compromise: The surface area under tension is large enough to give meaningful results, while the required specimen size is at the same time small enough to enable the specimens to be inspected by microscopy. The microscope used limited the short edge length of specimens to 200 mm. Therefore, 200 mm × 200 mm square specimens were used for the annealed glass test series (Section 7.3). 200 mm × 200 mm specimens are, however, too small to be tempered. This is why 300 mm × 200 mm rectangular specimens had to be used in the test series comprising heat treated glass (Section 7.4).

Table 7.1 compares the test setup and the specimen geometry of European standard tests with those of the present study.

Table 7.1: Comparison of coaxial double ring test geometries.

Test setup	Standard	Loading ring radius (mm)	Reaction ring radius (mm)	Tested area* (mm^2)	Specimen edge length (mm)
EN CDR R45	EN 1288-5:2000	9	45	254	100 × 100 (±2)
EN CDR R400	EN 1288-2:2000	300 ± 1	400 ± 1	240 000[†]	1 000 × 1 000 (±4)
present study		25.5	63.5	2 043	200 × 200 (Section 7.3)
					300 × 200 (Section 7.4)

* This is the surface area under uniform, equibiaxial tension = the area inside the loading ring (exception see [†]).
[†] This is the value from the standard. It does *not* correspond to the area within the load ring ($282\,743\,mm^2$).

There is already some experience with the 51 mm/127 mm ring geometry. Simiu et al. (1984), for instance, used the same loading ring dimension and Dalgliesh and Taylor (1990) and Overend (2002) used the same loading ring and reaction ring dimensions. The latter two performed the tests with direct steel-on-glass contact because this was found to give better results (less variance) by Dalgliesh and Taylor (1990). This configuration was, therefore, used for the present study.

Figure 7.2 shows photos of the loading and reaction rings. Both are made of steel S 235. A simple schematic representation of the test setup is shown in Figure 7.3. For a more detailed fabrication drawing, see Appendix C.

The coaxial double ring test jigs were installed in a hydraulic 200 kN universal testing machine. For strain measurements, 120Ω HBM strain gauge rosettes with three stacked measuring grids (0°/45°/90°, type 1-RY91-1.5/120) were glued to the specimen's tension face. For data acquisition, a HBM MGCplus data acquisition system and the HBM Catman v3.1 software package were used.

7.2. TEST SETUP AND MEASUREMENTS

Figure 7.2: Coaxial double ring test setup: loading ring (left) and reaction ring (right).

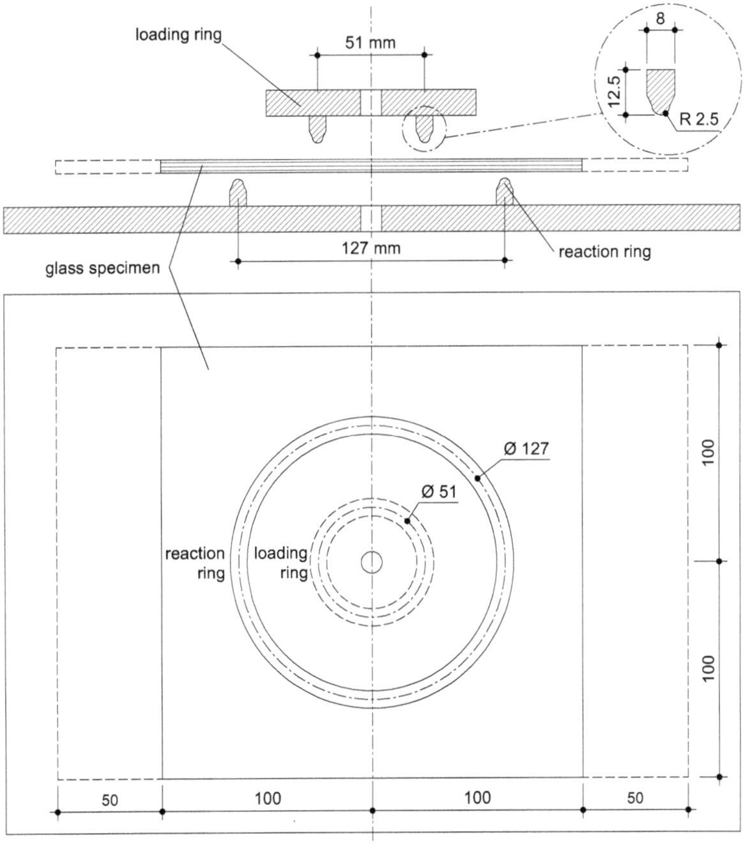

Figure 7.3: Schematic representation of the coaxial double ring test setup used (scale: 1:2.5).

Verification of the test setup's behaviour

The particularities of the test setup require the following questions to be verified analytically or numerically:

- Is there a significant difference in the mechanical behaviour of square and non-square specimens?
- What is the influence of the unknown amount of friction between the steel rings and the glass?

The test setup was modelled using the commercial finite element analysis software Abaqus (ABAQUS 2004). The specimen geometry was 300 × 200 mm or 200 × 200 mm. The glass thickness was 5.90 mm (mean of the measurements). A general purpose, 4-node, quadrilateral shell element with reduced integration (S4R) was used. Figure 7.4 shows the geometry and the element mesh for the square specimen. The glass is modelled as a linear elastic material with $E = 74\,\text{GPa}$[1] and $\nu = 0.23$. The loading and reaction rings were assumed to be rigid. The imposed loading ring displacement was applied in 200 steps of 0.01 mm each.

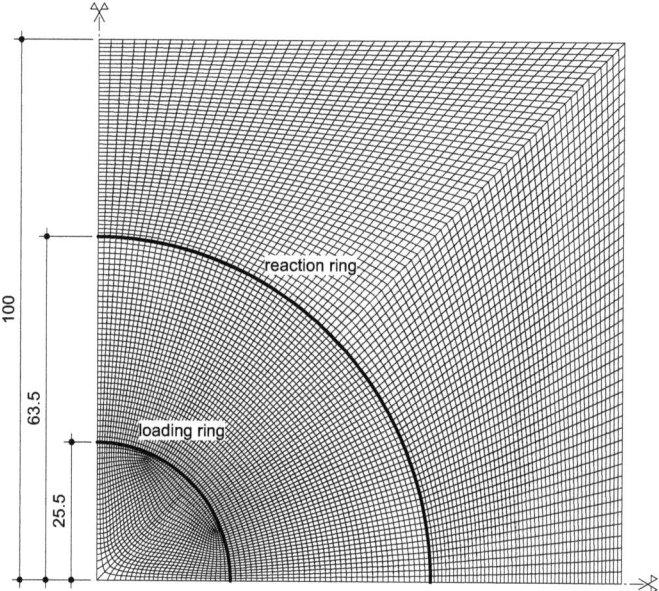

Figure 7.4: FE model of the coaxial double ring test setup (scale: 1:1).

Figure 7.5 compares the behaviour of the two specimen geometries. It can be seen that within the relevant stress range (< 250 MPa),

- glass of 6 mm nominal thickness is sufficient to keep deflections small compared to the plate thickness;
- the difference between the behaviour of non-square and square specimens is negligible.

In reality, there is an unknown amount of friction between steel and glass. Figure 7.6 compares the two extreme cases, *no friction* (free sliding) and *full friction* (no sliding). In the latter case, the stress below the loading ring increases. Furthermore, the load-stress curve slightly flattens out at very high loads. This is important because overestimation of the failure load would yield unsafe

---
[1] This is a likely value, while $E = 70\,\text{GPa}$ as cited in Table 2.5 is a nominal value for design.

7.2. TEST SETUP AND MEASUREMENTS

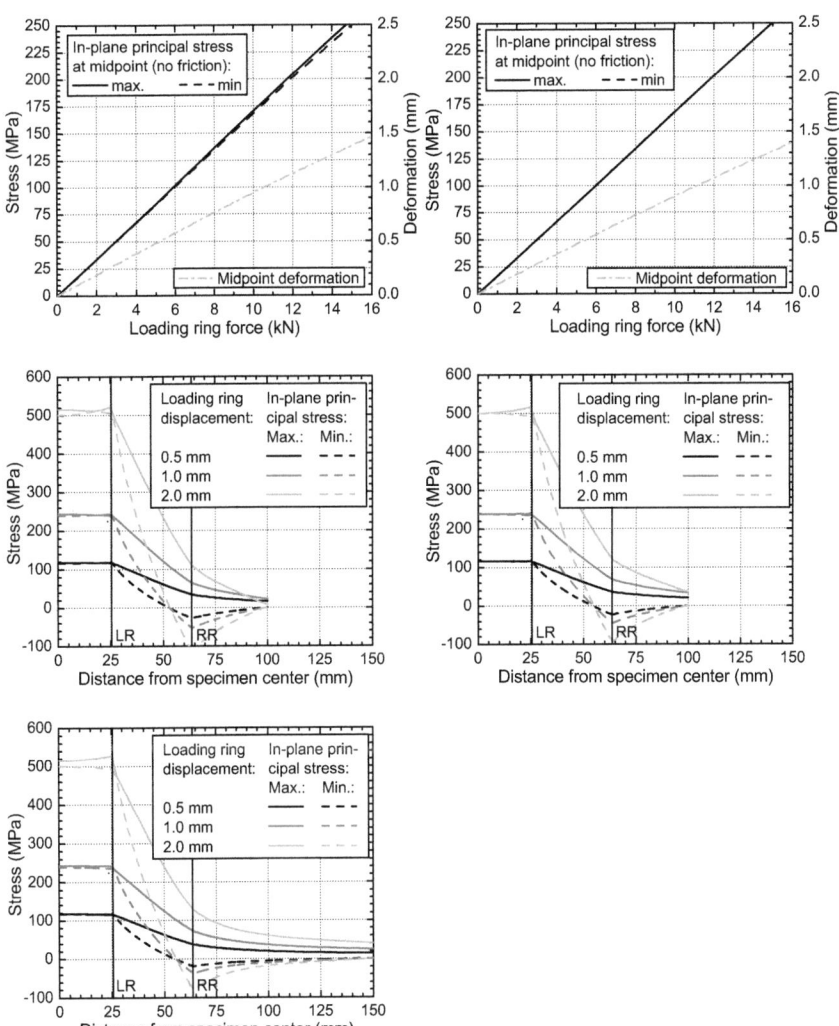

Figure 7.5: FE analysis of the test setup behaviour with 300 × 200 mm (left) and 200 × 200 mm (right) specimens. Negligible friction between steel rings and glass is assumed. LR = loading ring, RR = reaction ring.

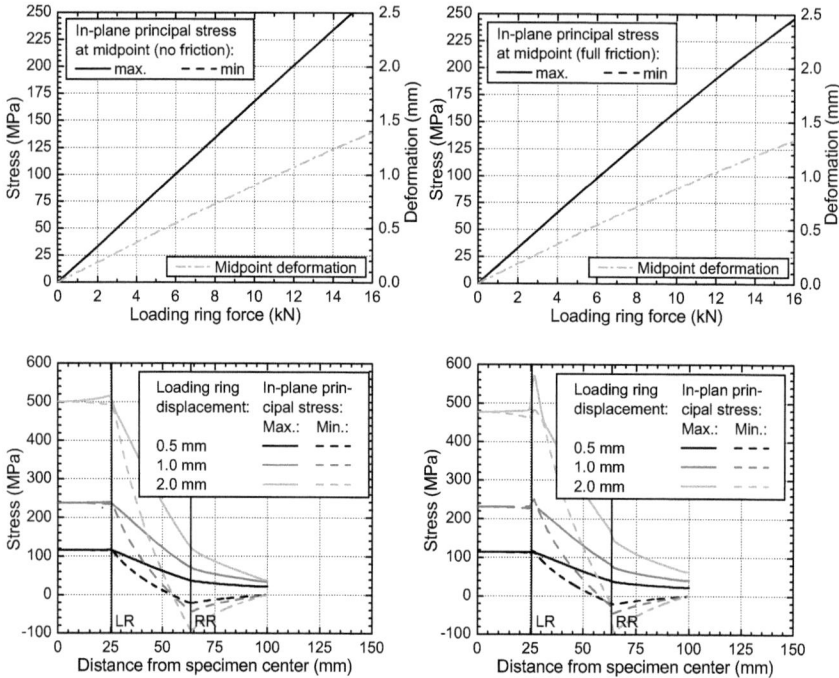

Figure 7.6: FE analysis of the 200 × 200 mm test setup behaviour without friction (left) and with full friction (right). LR = loading ring, RR = reaction ring.

resistance values. In order to find out what the *real* behaviour is, Figure 7.7 compares strain gauge measurements in the plate center with the above mentioned FE analysis results. The following can be seen:

- The measured minor and major in-plane principal stresses are not exactly equal. This is mainly due to the unevenness of heat treated glass specimens that is caused by the production process (cf. Section 2.3.2). The phenomenon is far less pronounced with measurements on annealed glass specimens. However, such measurements cannot provide data for loads above 5 kN.
- The quadratic polynomial fitted to the mean of the measured data is in very close agreement with the *full friction* model over the entire stress range. This polynomial ($\sigma_{\text{quadfit}} = a \cdot Q^2 + b \cdot Q$ with $a = -0.0570773\,\text{N}^{-1}\text{m}^{-2}$ and $b = 16510.6\,\text{m}^{-2}$) was, therefore, used as the nominal load-stress relationship when analysing test results.[2]

As the real behaviour is close to the full friction case, some stress concentration below the loading ring may occur (cf. Figure 7.6). To prevent negative impact on the accuracy of test results, specimens with failure origins below the loading ring are not considered.

Residual surface stress measurement

A crack's inert strength is its crack opening stress at failure (cf. Equations (4.10) and (4.12)):

$$\sigma_c(t_f) = \frac{K_{Ic}}{Y \cdot \sqrt{\pi \cdot a(t_f)}} = \sigma_E(t_f) + \sigma_r \qquad (7.1)$$

---
[2] For stresses $\leq 200\,\text{MPa}$, the linear relationship $\sigma_{\text{linfit}} = c \cdot Q$ with $c = 16003.5\,\text{m}^{-2}$ would also provide sufficient accuracy.

7.2. TEST SETUP AND MEASUREMENTS

Figure 7.7:
Surface stress in the center of the specimen: measured data, FE models, polynomial fit. IPP = in-plane principal stress, FEA = finite element analysis.

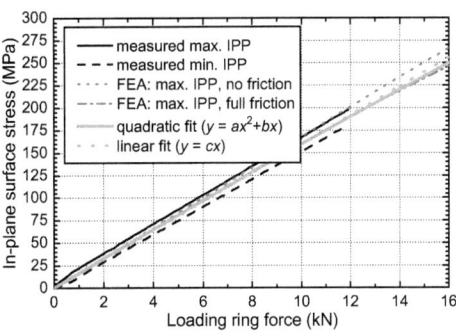

The depth of a flaw that is calculated from its measured strength using Equation (7.1) will be called *effective nominal flaw depth*[1] hereafter.

Strain gauges and the FE model described in Section 7.2 provided the surface stress due to loading σ_E. To determine the crack opening stress $\sigma_E + \sigma_r$, the residual stress σ_r was required. Because residual stresses vary considerably among specimens (cf. Section 6.6), tests on heat treated glass can only provide meaningful results if the residual stress of all specimens is measured prior to testing.

For the present investigations, the residual stress measurements were performed using an LDSR (differential stress refractometer) from Gaertner Scientific Corporation (Illinois, USA). This apparatus basically consists of a total internal reflection prism, a calibrated telescope, a light source and a mirror. It allows measurement of the birefringence[3] in the glass surface through the technique of total internal reflection made possible by the prism that is placed in optical contact with the glass surface. The magnitude of the surface stress is directly proportional to the birefringence and can thus be obtained from a calibrated conversion table (GaertnerCorp 2002).

LDSR measurements can only be performed on the tin side of the glass. The variation within the loading ring area is less than the measurement uncertainty, which is why all measurements are taken at the center of the specimens only. Even annealed glass has some small residual surface compression stress. However, such low stresses cannot be measured reliably with the LDSR.

Optical flaw depth measurement

The depth of surface flaws was measured optically using a confocal microscope. This type of microscope is able to exclude most of the light that is not from the microscope's focal plane. The image has less haze and better contrast than seen through a conventional microscope and represents a very thin cross-section of the specimen.[4] This allows flaw depths to be measured in a very straightforward way: The operator focuses first on the glass surface, then on the bottom of the crack. The difference in the focal distance between the two points is measured using a calibrated digital micrometer. This quantity will be called *optical flaw depth*[1] hereafter. The error involved in the optical flaw depth measurement is less than 2%.[5]

---
[3] This is the difference in refractive indexes for light rays oscillating perpendicular and parallel to the plane of incidence.
[4] For details on confocal microscopy, see e.g. Semwogerere and Weeks (2005).
[5] This was estimated using a glass specimen with a known thickness of 5.81 mm. The thicknesses measured using the microscope range from 3.841 mm to 3.880 mm which correspond, considering a coefficient of refraction of 1.52 (see Table 2.5), to 5.80 mm to 5.90 mm.

7.3 Creating near-inert conditions in laboratory tests

Testing procedure

As discussed in Section 6.7, laboratory testing at ambient conditions has notable drawbacks. Strength at ambient conditions depends on the surface condition *and* on the crack velocity parameters. As the latter is strongly stress rate dependent, test results obtained in conditions in which no subcritical crack growth occurs (inert conditions) would be more accurate and reliable. Considering the stress corrosion mechanism discussed in Section 4.1, inert conditions can be achieved in various ways:

1. Testing in a vacuum or in a completely dry environment.
2. Testing in a normal environment with a hermetic coating.
3. Testing in a normal environment at very rapid stress rates.
4. Testing at a sufficiently low temperature, at which the kinetics of environmentally induced reactions are arrested.

For simplicity, any test condition enabling inert strength measurements will be called *inert testing* hereafter, irrespective of whether testing is effectively done in an inert environment or whether an equivalent result is otherwise obtained.

Not all possibilities outlined above are equally suitable for structural applications. Options 1. and 4. are difficult and expensive, especially for full-scale testing on large specimens. Options 2. and 3., in contrast, are comparatively simple and inexpensive provided that the conditions do not need to be fulfilled perfectly. This is the idea behind the present experiments. They serve to verify whether the inert strength can be measured by combining a nearly-hermetic coating and a quite rapid stress rate. The latter reduces the effect of the former's imperfection and vice versa. The specimens were prepared and tested as follows:

1. **Drying.** The specimens were dried in an oven at $100 \pm 5\,°C$ for 48 ± 6 hours. The humidity in the oven was maintained below 5% RH by a high performance molecular sieve desiccant[6].
2. **Hermetic coating.** To achieve a hermetic coating, a silicone grease[7] was applied to the tension face of the specimens. It is highly hydrophobic, impermeable and its viscosity is high enough to ensure that the coating remained intact during handling and testing.
3. **Adoption.** Specimens were kept at ambient conditions for 2 hours to allow them to adopt ambient temperature'.
4. **Destructive testing.** The specimens were loaded to failure using the coaxial double ring test setup described in Section 7.2.

Experiments and results

Square specimens with an edge length of 200 mm and a nominal thickness of 6 mm were tested using the coaxial double ring test setup described in Section 7.2. All specimens were made of as-received, annealed soda lime silica float glass. The testing programme is summarized in Table 7.8.

The results are summarized in Figure 7.9. Detailed data are provided in Appendix C. The choice of the estimator for the empirical probability that is used to plot the experimental data is discussed in Section E.2.

First, two test series (AS, AF) were conducted at ambient conditions at different stress rates to obtain some reference data. Then, a first series of coated specimens (IS) was tested at a low stress rate. It can be seen in Figure 7.9 that the result was unsatisfactory. While the failure stresses are clearly higher than without a coating at the same stress rate, they lie below the results of uncoated specimens tested at a rapid stress rate. This means that subcritical crack growth has only been prevented to a limited extent. Increasing the stress rate by a factor 100 (series IF) yielded significantly better results.

---

[6] Supplier: Zeochem AG, 8707 Uetikon am See, Switzerland.
[7] Product: Rhodorsil Pat 4. Supplier: Silitech AG, 3000 Bern, Switzerland.

7.4. DEEP CLOSE-TO-REALITY SURFACE FLAWS

Table 7.8:
Test series with as-received annealed glass specimens.

Series	Specimens	Average eff. load rate (kN/s)	Average eff. stress rate (MPa/s)	Test conditions
AS	10	0.0130	0.21	ambient
AF	10	1.32	21.2	ambient
IS	10	0.0128	0.21	dried, coated
IF	10	1.35	21.6	dried, coated
IV	4	2.46	39.4	dried, coated

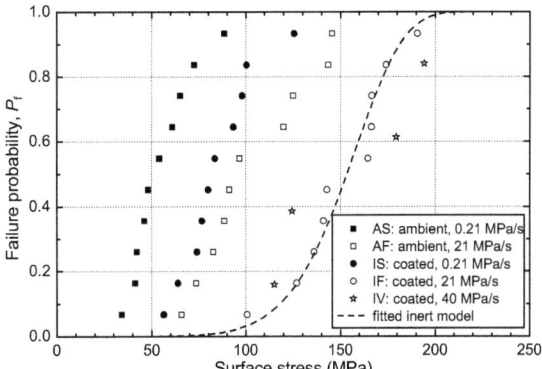

Figure 7.9:
Test results of as-received annealed glass specimens.

There may, however, still be some crack growth. In order to verify this, the stress rate was doubled for an additional series (IV). The fact that the failure stress did not increase significantly (Figure 7.9) suggests that series IF was already very close to ideally inert conditions.

This is convenient because common testing equipment tends to cause various problems at very high stress rates (hydraulics, machine control, data acquisition). These problems can negatively affect accuracy and even yield unsafe results, which is why very high stress rates should generally be avoided. If subcritical crack growth is not prevented altogether during near-inert tests, the results are conservative. This is a major advantage over ambient testing, in which overestimation of the crack growth during the tests can yield unsafe results.

Fitting the inert strength model to the as-received glass strength data (dashed line in Figure 7.9) yields the following surface condition parameters: $\theta_0 = 67.6$ MPa and $m_0 = 7.2$. This is in close agreement with the parameters obtained in Section 6.5 from very high stress rate testing at ambient conditions ($\theta_0 = 63$ MPa and $m_0 = 8.1$).

7.4 Deep close-to-reality surface flaws

7.4.1 Testing procedure

Surface flaws created on specimens for glass testing have to meet two contradictory requirements:

1. They should be as similar as possible to the surface damage that structural glass elements are likely to undergo in in-service conditions. This includes accidental damage (e. g. due to handling, cleaning, impact of vehicles, tools falling down or impact of heavy wind-borne debris) as well as intentional damage (vandalism).

2. They should be as reproducible as possible.

106 CHAPTER 7. EXPERIMENTAL INVESTIGATIONS

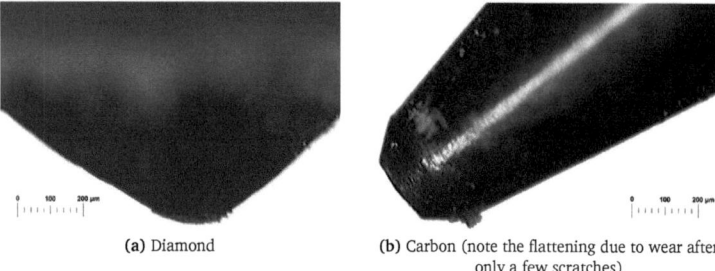

(a) Diamond

(b) Carbon (note the flattening due to wear after only a few scratches)

Figure 7.10: Tips used for surface scratching.

Figure 7.11: Surface scratching device.

In order to achieve an optimal compromise, long surface cracks are created by applying a reasonably constant force to a thin, sharp tip.[8] Two types of scratching tips are chosen for a more detailed examination: a 0.33 carat dressing diamond and a sharp carbon tip (Figure 7.10).

To be able to apply a constant force to the scratching tip, the surface scratching device shown in Figure 7.11 was developed. A steel plunger holds the diamond or carbon tip. A casing guide ensures that the plunger is positioned exactly perpendicular to the glass plate. Ball bearings are used to minimize the sliding friction between the plunger and the guide. The plunger can be loaded with steel blocks of known weight and creates a constant contact pressure between the scratching tip and the specimen. Figure 7.12 shows the geometry of the specimen, the steel rings and the surface scratches.

Preliminary tests were conducted on annealed glass specimens in order to verify the reproducibility of the flaw depth and to select the more suitable scratching tip. The flaw depth was measured optically. The tests allow the following conclusions to be drawn:

- The carbon tip is unsuited for the task at hand. Its originally sharp tip wears very rapidly, so that it would have to be replaced frequently to obtain a sufficiently constant crack geometry.

- The diamond behaves much better in terms of wear. Furthermore, its relatively large opening angle causes some widening of the scratch, which is an effect that is likely to happen with objects commonly used by vandals (e. g. diamond rings).

---

[8] Some preliminary tests were also done with commonly used glass cutting devices. But while the scratches are more reproducible (requirement 2), they are not very similar to in-service surface damage (requirement 1).

7.4. DEEP CLOSE-TO-REALITY SURFACE FLAWS

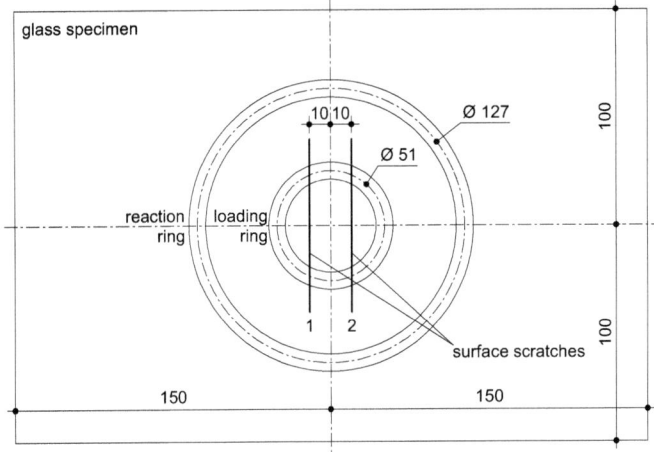

Figure 7.12: Geometry of the specimen, the steel rings and the surface scratches (scale: 1:2.5).

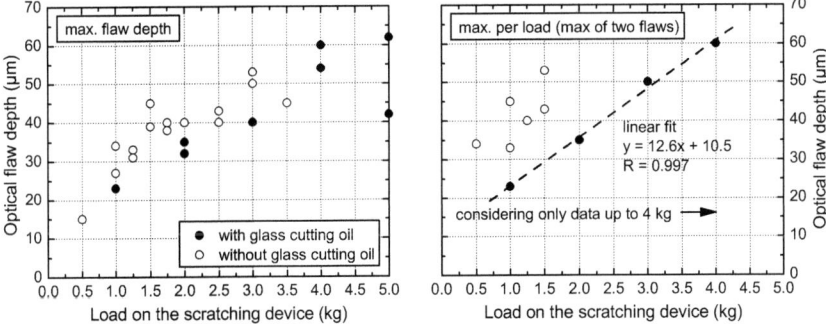

Figure 7.13: Optical flaw depth as a function of the scratching device load (diamond tip).

- In dry diamond on glass scratching, the regularity of the surface flaws is problematic. The scatter of the flaw depths is large, the flaw geometry uneven and the diamond becomes stuck with scratching device loads above 3 kg. If a glass cutting oil is used, depth and geometry are more uniform. The diamond glides more smoothly over the surface, which allows higher scratching device loads to be applied (Figure 7.13).

Based on this experience, the following testing procedure was used for the main test series:

1. **Surface scratching** as described above, with scratching device loads ranging from 0.5 kg to 3.5 kg. The evaporating glass cutting oil Glasol GB[9] was applied to the surface before scratching.
2. **Optical crack depth measurement** as described in Section 7.2.
3. **Destructive testing** using the coaxial double ring test setup described in Section 7.2. For inert strength measurements, the drying and hermetic coating procedure described in Section 7.3 was used. For ambient strength measurement, no coating was applied.

---
[9] This oil does not leave a residue on the glass surface after evaporation and does not attack glass coatings.

7.4.2 Experiments and results

The above-mentioned testing procedure was applied to annealed, heat strengthened and fully tempered soda lime silica float glass specimens. The specimens' dimensions were 300 mm × 200 mm and the nominal glass thickness was 6 mm. Table 7.14 gives an overview of the testing programme. For tables with detailed test results, see Appendix C.

Series	Average effective load rate (kN/s)	Average effective stress rate (MPa/s)	Glass type
SI.A	1.148	18.6	annealed
SI.H	1.293	20.2	heat strengthened
SI.F	1.322	19.9	fully tempered

Table 7.14: Test series with dry and silicone-coated specimens containing deep surface flaws.

Before examining the strength of surface flaws, the specimens were used to assess the influence of the residual stress on the sensitivity of the glass surface to contact damage. Figure 7.15 shows the optical flaw depth as a function of the scratching device load. The most evident observation from this figure is that the scatter of the data is very large. As far as the optical flaw depth is concerned, there is no clear evidence of the residual stress level having an influence on the sensitivity of the glass surface to contact damage.[10] With a scratching device load of 3.5 kg, the maximum and the mean optical flaw depths are generally smaller that with 2.5 kg. The reason for this is that the scratches become less smooth and 'clean' as the scratching device load increases because the diamond tip tends to become stuck. This behaviour sets an upper limit to the flaw depth that can be introduced for testing. Furthermore, it suggests that there may be an upper limit of the flaw depth that is likely to be introduced by accidental or intentional scratching in in-service conditions.

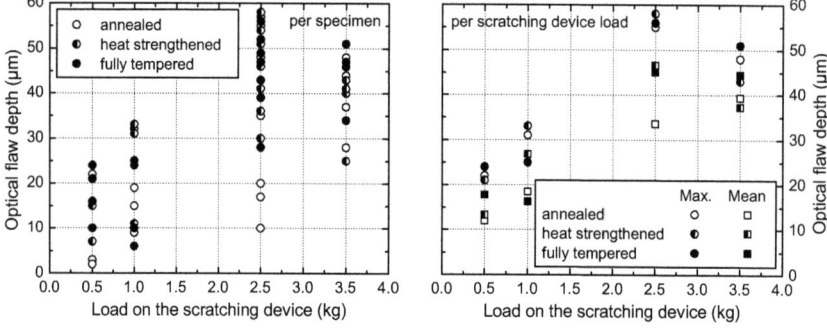

Figure 7.15: Measured optical flaw depths.

Figure 7.16 shows the relationship between the measured inert strength (cf. Equation (7.1)) and the optical flaw depth. Detailed data are given in Appendix C. The inert strength model (dashed line) is based on $Y = 1.12$. Figure 7.17 provides an alternative view, additionally showing the dependence on the scratching device load. Figure 7.18 compares the optical flaw depth with the effective nominal flaw depth. In the case of a perfect correlation, all data points would lie on the dashed diagonal line.

Although the scatter of the data is very large, the three figures show an interesting glass type-dependence of the behaviour of deep flaws. The inert strength of flaws in fully tempered glass follows the model. In heat strengthened glass, it lies somewhat below the model, but follows the model's trend. In annealed glass, the inert strength is clearly below the model and is actually not a function of the optical flaw depth.

---

[10] There is a difference with respect to *strength*, see below.

7.4. DEEP CLOSE-TO-REALITY SURFACE FLAWS

Figure 7.16:
Inert strength of flawed specimens as a function of the optical flaw depth. (The vertical arrow shows the inert strength range of as-received annealed glass from Figure 7.9.)

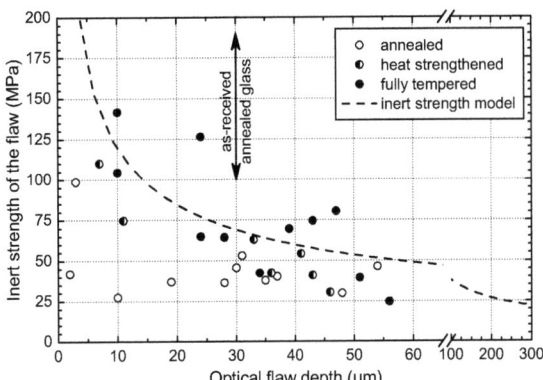

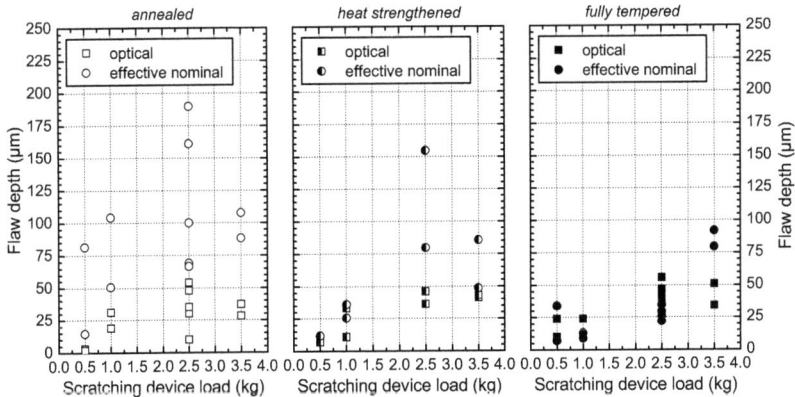

Figure 7.17: Optical flaw depth and effective nominal flaw depth as a function of the scratching device load and the glass type.

Figure 7.18:
Correlation between optical flaw depth and effective nominal flaw depth.

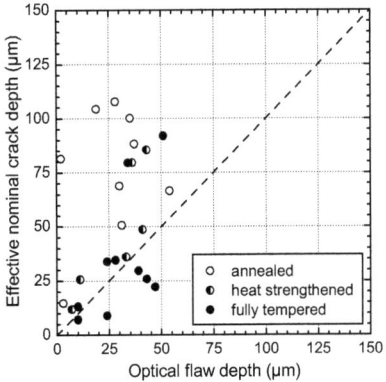

In an attempt to explain this behaviour, scratched specimens were broken perpendicularly to the scratches and the cut ends were investigated under the microscope. A typical case is shown in Figure 7.19. It can be seen that the fractured glass zone around a surface scratch is significantly deeper than the open, visible depth (optical flaw depth). This means that the glass fracture caused by the scratching tip extends significantly beyond the zone that is in direct contact with the tip. The extension is hindered by compressive stresses. This causes the difference between the optical and the effective nominal flaw depth to decrease as the residual stress level increases. This explains the differences among glass types that were observed in Figures 7.16, 7.17 and 7.18.

In conclusion, the residual stress level has an influence on the sensitivity of the glass surface to contact damage. The higher the residual compressive stress, the lower the effective nominal flaw depth for a given scratching force is.

Equation (7.1) should allow the geometry factor Y to be calculated if the inert strength of specimens with cracks of known depth is determined. In view of the above findings, however, this turns out to be unrealistic because the effective nominal flaw depth cannot be measured using an optical microscope. This clearly confines the usefulness of optical flaw depth measurements.

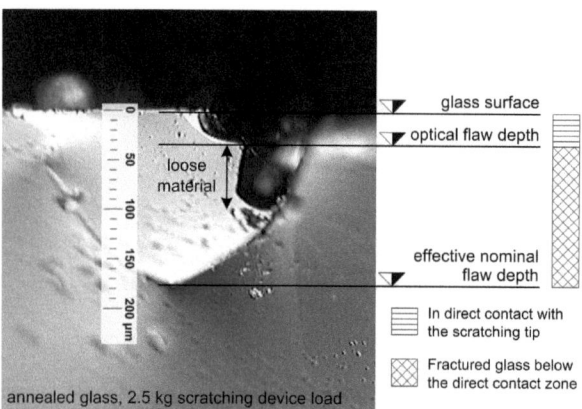

Figure 7.19: Optical flaw depth versus effective nominal flaw depth (the loose material fell out when breaking the specimen).

7.4.3 Is hermetic coating indispensable for deep flaw testing?

The presence of a hermetic coating is expected to have a very limited effect on the strength when specimens with deep flaws are tested at high stress rates (see Section 8.3.1). If this can be confirmed by experiments, surface coating would not be required in such tests. To verify this, some of the scratched specimens are tested at ambient conditions without surface coating. Table 7.20 gives an overview of the testing programme. For tables with detailed results, see Appendix C.

Series	Average effective load rate (kN/s)	Average effective stress rate (MPa/s)	Glass type
SA.A	1.299	21.1	annealed
SA.H	1.388	21.8	heat strengthened
SA.F	1.407	21.4	fully tempered

Table 7.20: Test series with uncoated specimens containing deep surface flaws.

Figure 7.21 shows a comparison of measured crack opening stresses at failure with and without surface coating. Indeed, there is only a very faint tendency of non-coated specimens towards lower strength. This means that in the case of specimens with deep flaws, tests without hermetic coating

7.5. VISUAL DETECTABILITY OF DEEP SURFACE FLAWS

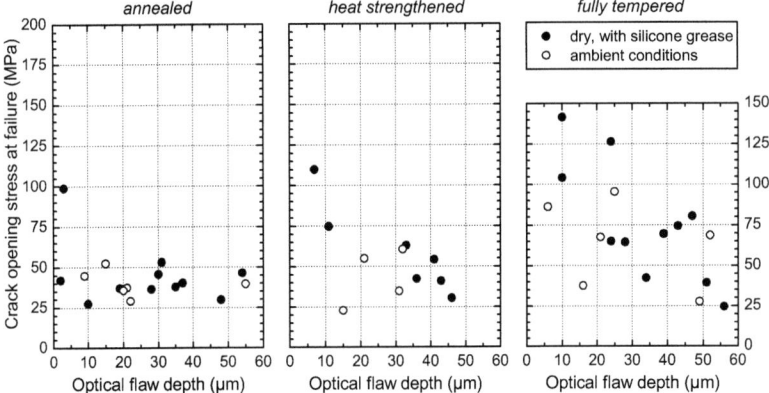

Figure 7.21: Comparison of the strength of deep flaws at ambient and at inert conditions.

yield only very slightly more conservative results. Because of the longer time to failure and the small crack depth, the same is *not* true for as-received glass specimens (cf. Section 7.3).

Tests without coating are less expensive, faster and avoid uncertainty related to crack healing during the drying period.

7.5 Visual detectability of deep surface flaws

The preceding sections have shown the large effect of deep surface flaws on the resistance of structural glass elements. It is current practice to inspect glass facades visually for damage after their installation. Such a visual inspection is usually done by the building owner or his representative. It normally consists of looking at the glass from a distance of 3 m and at a viewing angle of 90° with respect to the glass surface. Glasses containing defects that are visible using this procedure are replaced.

At least in the case of glass elements that are installed in locations in which they are permanently and safely protected from surface damage, such inspections could allow for less conservative design. If the depth of the deepest flaw that may be overlooked when viewing the glass from 3 m in the inspection was known, this information could be used to define the maximum crack depth that must be considered for design (design crack depth).

To this end, the experiments described below were conducted. They were designed to assess the relationship between depth and visual detectability of flaws as well as to quantify the maximum depth of flaws that may be overlooked. However the following must be considered:

- If glass elements are potentially exposed to surface damage during their service life, visual inspections have to be repeated regularly. Even then, glass elements may contain flaws above the detectability limit for a time period of up to the inspection interval.

- Visual inspection of installed glasses obviously cannot provide information on the surface condition of hidden regions such as concealed holes and edges. These must be inspected before installation and damage during installation has to be prevented by appropriate protection and careful handling. Once in place, damage to hidden regions of a glass element can usually be prevented by good practice detailing.

Testing procedure

The testing procedure was as follows:

- Glass specimens containing surface flaws of variable depth (including no visible flaws at all) were installed at eye level.

- Four different people inspected the specimens for visible flaws, looking at them perpendicularly from a distance of 3 m.

The detectability of surface flaws depends strongly on the background behind the glass and the reflection on the glass (Figure 7.22). In daylight, most reflection generally occurs when looking at a facade from the outside. When looking out from an interior space or at an element inside a building, there is much less reflection. To take this into account, all testing was done twice: once looking inwards towards a rather dark interior space, once looking outwards to a sunny natural background. The detectability of a total of 56 flaws was assessed this way.

Experiments and Results

Figure 7.23 shows the experimentally determined probabilities of visual detection. As expected, there is no clear evidence of a correlation between optical flaw depth and probability of detection. The visibility of a flaw depends much more on its shape than on its depth. It can be seen, however, that very deep flaws were always detected. This allows for the definition of a lower visual detectability limit $a_{VDL} = 40\,\mu m$. This corresponds to the maximum optical depth of a surface flaw that might

Figure 7.22: The influence of the viewpoint on visual detectability: inwards view, high viewpoint (top); inwards view, low viewpoint (middle); outwards view (bottom).

7.5. VISUAL DETECTABILITY OF DEEP SURFACE FLAWS

be overlooked. The value must, however, be used with caution. Tools that are sharper than the diamond used in the present tests might create flaws that may be overlooked despite their greater depth. For application, it is advisable to define relevant surface damage hazard scenarios as part of a comprehensive risk analysis and to do detectability testing based on this damage.

Figure 7.24 shows the optical flaw depth and the visual detectability limit together with the effective nominal depth of the flaws that was determined by destructive testing (cf. Section 7.4). It can be seen that for *heat strengthened and fully tempered glass*, the effective nominal flaw depth of non-detectable flaws lies below about 100 μm. For *annealed glass*, however, the effective nominal flaw depth of non-detectable flaws may reach about 200 μm.

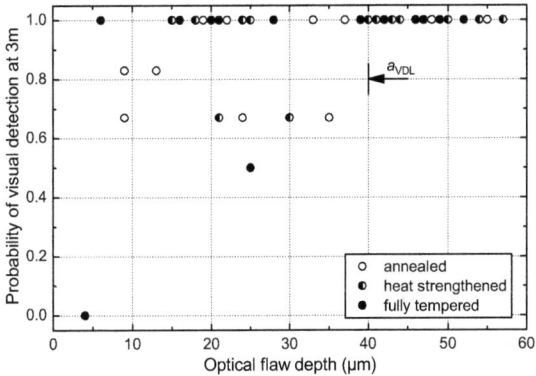

Figure 7.23: Probability of visual detection of surface flaws (experimental results).

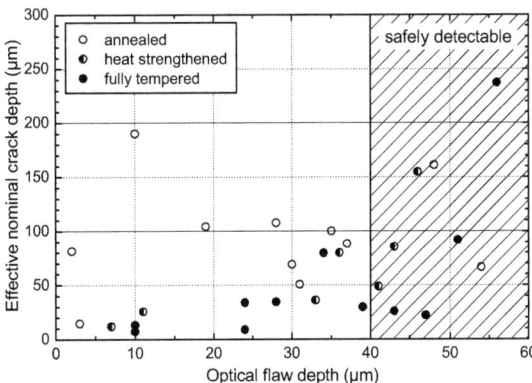

Figure 7.24: Visual detectability and effective nominal flaw depth.

7.6 Summary and Conclusions

⇨ **Improvement of existing testing procedures.**
- It has been shown that near-inert conditions can be created in laboratory tests using inexpensive and generally available equipment. The proposed testing procedure prevents crack growth by a combination of drying, nearly-hermetic coating (covering the surface with a silicone grease) and a relatively rapid stress rate. If, for some reason, subcritical crack growth should not be prevented altogether, the results are conservative. This is a major advantage over testing at ambient conditions, in which overestimation of crack growth during the tests can yield unsafe results.
- Specimens with deep surface flaws may also be tested at a high stress rate but without drying and hermetic surface coating. Because of the short time to failure, the results are only slightly more conservative compared to those from tests with drying and coating. Therefore, this is a less expensive and faster alternative. This procedure is, however, not suitable for as-received glass specimens.
- The developed surface scratching device allows for the creation of deep scratches on glass specimens that are fairly reproducible and similar to those expected in accidental damage or vandalism hazard scenarios (design flaws).

⇨ **Mechanical behaviour of deep close-to-reality surface flaws.**
- The scatter of the strength of deep surface flaws is extremely high.
- It has been seen that the fractured glass zone around a surface scratch is significantly deeper than the open, visible depth (optical flaw depth). The effective nominal flaw depth that governs strength is, therefore, significantly deeper than the optically measured flaw depth.
- Residual compressive stress hinders fraction of the glass beyond the zone that is in direct contact with the scratching tip. This is the reason why the residual surface stress level influences the effective nominal flaw depth, although it has no significant effect on the optical flaw depth. The strength reduction caused by surface damage decreases as the residual compressive surface stress increases.

⇨ **Surface condition parameters.** The surface condition parameters of as-received glass that have been determined from near-inert tests in the present chapter are in close agreement with the parameters obtained in Section 6.5 from very high stress rate testing at ambient conditions.

⇨ **Geometry factor of deep flaws.** The geometry factor of deep close-to-reality surface flaws cannot be determined by combining optical flaw depth measurement with destructive testing because the effective nominal flaw depth that governs strength cannot be measured using an optical microscope.

⇨ **Detectability.** The maximum optical depth of surface flaws that may be overlooked during the standard visual inspection from a distance of 3 m is about 40 µm. The effective nominal depth of such flaws lies below about 100 µm for heat strengthened and fully tempered glass and below about 200 µm for annealed glass. These experimental results must, however, be used with caution. The detectability of surface flaws depends strongly on the background and on reflections on the glass. Furthermore, tools that are sharper than the diamond used in the present tests might create flaws that may be overlooked despite their greater depth.

Chapter

8

Structural Design of Glass Elements

In this chapter, the developments and findings of all preceding chapters are deployed in order to provide recommendations for the structural design of glass elements and for the laboratory testing required within the design process. To this end, the chapter commences by discussing key aspects related to the use of the lifetime prediction model for structural design. On this basis, recommendations are then developed. The chapter closes with a short presentation of the computer software that was developed to facilitate the application of the recommendations and to enable the lifetime prediction model to be used efficiently in research and practice.

8.1 Introduction

It is pertinent to recall and summarize the key aspects that govern the structural behaviour of glass elements and cause their structural design to differ considerably from the design of elements made of more common construction materials:

- **Surface condition.** The resistance of a glass element is very sensitive to the flaws on its surface. In addition to standard hazard scenarios, surface damage hazard scenarios should, therefore, be considered for design. Such hazard scenarios represent factors that cause severe surface damage such as accidental impact, vandalism, or heavy wind-borne debris.

- **Subcritical crack growth.** Stress corrosion causes flaws on a glass surface to grow subcritically. The resistance of a loaded glass element therefore decreases with time, even if it is exposed only to static loads. The growth of a surface flaw depends on the properties of the flaw and the glass (initial crack depth, geometry factor, crack growth threshold), the crack opening stress history that the flaw is exposed to (which is a function of the element's geometry, the support conditions and the loading), and on the relationship between crack velocity and stress intensity (represented by the crack velocity parameters). The strong dependence of the crack velocity parameters on external influences such as temperature, humidity and the loading rate make accurate predictive modelling difficult. For structural design, a conservative estimate of the crack velocity parameters can be used. Experiments should be performed at inert conditions in order to prevent subcritical crack growth and its barely predictable effect on the results.

- **Surface decompression.** In surface regions that are not decompressed, there is no positive crack opening stress and consequently no subcritical crack growth. This is of particular importance for heat treated glass, in which surface decompression only occurs if the tensile stress due to loading exceeds the residual surface compression stress.

Subcritical crack growth and surface decompression were extensively discussed in Chapter 4 and the lifetime prediction model takes these aspects into account. What clearly deserves further consideration is how to model a glass element's surface condition. The lifetime prediction model offers two alternatives (cf. Chapter 4): a *single surface flaw (SSF)* and a *random surface flaw population (RSFP)*. It is essential to know which one to use and when. After explaining the structural design process (Section 8.2), the characteristics and particularities of these two surface condition models will therefore be discussed in Section 8.3. On this basis, recommendations for design and testing can then be given in Section 8.4. Finally, Section 8.5 presents computer software that was developed to facilitate the application of the recommendations and to enable the lifetime prediction model to be used efficiently in research and practice.

8.2 The structural design process

Predictive modelling is the process of using a model to predict the probability of failure of a glass element that is exposed to a given action history. *Structural design* is based on predictive modelling, but goes one step further. It is the iterative process of selecting a glass element that meets a set of performance requirements that depends on the specific application. Common requirements for structural glass elements relate to aspects such as deformation, vibration, usability, aesthetics, acoustic or optical performance, and, of course, load bearing capacity, which is what the present study concentrates on. Structural design is based on predictive modelling. But while the failure probability is an outcome of predictive modelling, it is a target value for design.

Figure 8.1 shows the structural design process of glass elements using the lifetime prediction model. It can be seen that three input models are required:

- **Material model.** The material model includes material parameters (K_{Ic}, σ_r), crack velocity parameters (v_0, n) and a representation of the surface condition. The latter may either be a *single surface flaw (SSF)* that is characterized by its initial depth (a_i) and a geometry factor (Y) or a *random surface flaw population (RSFP)* that is characterized by surface condition parameters (θ_0, m_0) and a geometry factor (Y).

- **Structural model.** A structural element is characterized by its geometry, its geometrical imperfections, the support conditions and the loading conditions. A non-linear finite element analysis provides the action/stress relationship (e. g. load/stress, displacement/stress, temperature/stress), namely the in-plane principal stresses on all faces of all finite elements as a function of the action intensity[†].

- **Action model.** Action models may take on many forms of varying complexity. A minimal definition consists of a probability distribution representing the action intensity and a characterization of the behaviour with respect to time (e. g. the duration of a realization[†] and the number of realizations per reference time period). An action history generator calculates all realizations and superposes all actions. The final output is an action intensity history.

The output of these models is used in the lifetime prediction model in order to calculate the probability of failure. A structural element is acceptable if its probability of failure during the service life is less than or equal to the *target failure probability* $P_{f,t}$ (random surface flaw population) or if the design flaw does not fail during the design life (single surface flaw).

8.3 Characteristics of the two surface condition models

8.3.1 Single surface flaw model

The surface condition of as-received glass can be characterized accurately by a random surface flaw population (RSFP), i. e. a large number of flaws of random depth, location and orientation (cf. Section 4.3 and Section 5.1). If, however, a glass element's surface contains a flaw (or a few flaws)

8.3. CHARACTERISTICS OF THE TWO SURFACE CONDITION MODELS

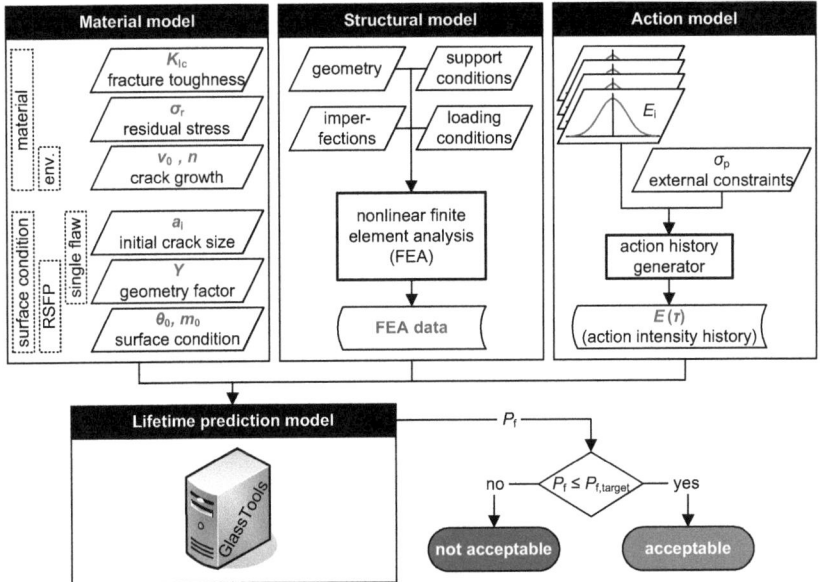

Figure 8.1: Structural design of glass elements with the lifetime prediction model.

that is substantially deeper than the many small flaws of the RSFP, its resistance will inevitably be governed by this deep flaw because it will fail first.

If the surface condition of a glass element can be represented by a single surface flaw, its lifetime can be predicted by simulating the growth of this flaw using the equations derived in Section 4.2. A glass element is acceptable if a design flaw does not fail during the service life when the element is exposed to the design action history.

In order to determine the stress history that the design flaw is exposed to, its location must be known. In some cases, e. g. bolt holes, it makes sense to choose a particular flaw location. In other cases, the location of the design flaw may be completely unknown. In such cases, it is safe to assume the flaw is located anywhere on the surface, i. e. to simulate crack growth using the 'worst' stress history (the one that causes the most crack growth) that exists on the element's surface.

Behaviour of a surface crack

Figure 8.2 quantitatively illustrates the behaviour of a surface crack by numerically solving Equation (4.20) for the failure stress σ_f. It shows the stress that causes failure when the stress is held constant over t_f for different initial crack depths. In contrast to what would be obtained from the simplified formulation in Equation (4.19), the failure stress rightly converges on the inert strength for very short loading times. The left graph in the figure assumes $v_0 = 6\,\text{mm/s}$, which was found to be a reasonable and conservative assumption for design purposes in Section 6.3 (fast crack growth). The right graph assumes $v_0 = 0.01\,\text{mm/s}$ (slow crack growth). This is the value that was estimated in Section 6.5 for constant stress rate testing at $2\,\text{MPa/s}$. The figure shows the following:

- The strength of cracks is strongly time-dependent.
- The long-term strength of cracks with an initial depth in the order of 100 μm or more is low.
- In laboratory conditions (right graph), deep cracks do not grow significantly if loaded for less than a few seconds. This is something that one can take advantage of when doing laboratory testing on specimens with deep surface flaws, see Section 8.4.2.

CHAPTER 8. STRUCTURAL DESIGN OF GLASS ELEMENTS

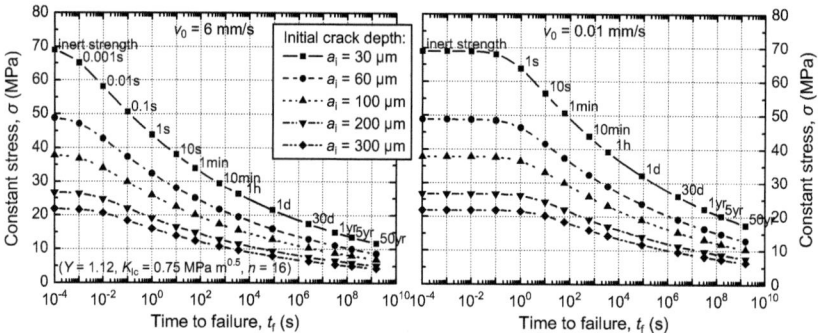

Figure 8.2: Strength of a crack as a function of time for fast (left) and slow (right) crack velocity.

Conclusions

In light of its derivation (Section 4.2) and the above considerations, the main characteristics of the single surface flaw model can be summarized as follows:

- This model is suitable for cases in which a glass element's failure is governed by one deep surface flaw or a few such flaws.

- It caters for arbitrary geometries and loading conditions, as long as sensible assumptions with regard to the location of the design crack and the crack opening stress history at this location can be made.

- Since the outcome of the model is a function of the conditions at the location of the design flaw(s) only, it is not influenced by the element's size or by biaxial or non-homogeneous stress fields.

- Because of the simple representation of the surface condition, the model is intuitive, easy to use and numerical modelling is simple and fast. Furthermore, no statistical representation of the surface condition is integrated into the model. This is an advantage compared to random surface flaw population-based modelling, because it enables the design crack depth to be sampled from any distribution function considered appropriate for the specific task at hand. Information from testing, inspection, proof loading or engineering judgement can easily be integrated into the model through an adapted, possibly discontinuous, distribution function or a cut-off.

8.3.2 Random surface flaw population model

With this approach, the surface condition of a glass element is represented by a random surface flaw population, i. e. by a large number of flaws whose number, location, orientation and depth are all represented by statistical distribution functions (cf. Section 4.3 and Section 5.1).

The lifetime of a glass element is predicted by simulating the growth of its surface flaw population when the element is exposed to the design action history using the equations derived in Section 4.3. An acceptable design is achieved when the probability of failure during the service life is less than or equal to the target failure probability.

In Section 5.1, the random surface flaw population was found to represent the surface condition of as-received glass well. It may, however, often be unrepresentative of in-service conditions, especially if deep surface flaws occur or if structural elements contain machining damage.

8.3. CHARACTERISTICS OF THE TWO SURFACE CONDITION MODELS

Problems associated with the definition of the target failure probability

As mentioned in Section 8.2, the failure probability is a target value for structural design. This is an important difference from predictive modelling. Accordingly, the design equation in Section 5.3.1 contains the *target failure probability* $P_{f,t}$ as an additional parameter. To keep the verification format simple, $P_{f,t}$ is taken into account within the definition of the reference inert strength $f_{0,\text{inert}}$. This quantity is, therefore, a function of the surface condition, as well as of the target failure probability (definition, see Equation (4.52)).

Defining an appropriate target failure probability is less straightforward than one may think and is therefore worth close consideration.

It was seen in Chapter 3 that current design methods are based on two rather different failure probabilities. While European design methods use $P_{f,t} = 0.0012$, North American design methods use $P_{f,t} = 0.008$. The latter value was chosen by the developers of the GFPM because it was found to give 'reasonable results'. The former value was chosen based on the European standard EN 1990:2002. According to this document, design resistance should be defined so that

$$P_{f,R_d} = \mathscr{P}(R \leq R_d) = \Phi(-0.8 \cdot \beta) \tag{8.1}$$

where R is the effective resistance, R_d the design resistance, β the target reliability index and Φ the cumulative distribution function of the standard normal distribution. Equation (8.1) implies a fixed design point, an approach that enables the design resistance to be defined independently of the statistical properties of the actions. This is an approximation and requires the definition of fixed influence factors, in EN 1990:2002 -0.8 for the resistance and $+0.7$ for actions, which may or may not be appropriate for glass. They were defined based on experience with non-glass structures (mainly steel and concrete structures). For standard buildings and a service life of 50 years, EN 1990:2002 specifies a target reliability index of $\beta = 3.8$, which yields the above-mentioned value of $P_{f,t} = 0.0012$.

The fact that one single glass element is currently designed with different target failure probabilities depending on the design method used gives rise to two questions: How important is the choice of the target failure probability, i. e. how sensitive is the design to the choice of the target failure probability? Why do different design methods adopt such different target failure probabilities?

To answer the first question, it is useful to look at a few graphs. Figure 8.3 and Figure 8.4 show the inert reference strength $f_{0,\text{inert}}$ and the corresponding initial crack depth $a_i(f_{0,\text{inert}})$ as functions of the surface condition parameters m_0 and θ_0 for the two target failure probabilities $P_{f,t}$ mentioned above. Figure 8.5 shows the dependence of $f_{0,\text{inert}}$ and $a_i(f_{0,\text{inert}})$ on $P_{f,t}$. The following can be seen from the figures:

- The reference inert strength $f_{0,\text{inert}}$, which is actually the design resistance, depends strongly on the surface condition parameter m_0 and the target failure probability $P_{f,t}$.

- The lower the target failure probability $P_{f,t}$ is, the more important the influence of m_0 becomes.

- For low $P_{f,t}$ and rather low but realistic m_0 values, the initial crack depth a_i that corresponds to the resulting reference inert strength becomes unrealistically high.

Additional insight can be gained by investigating how the target failure probability affects the design resistance that results from various test series and design methods. This information is shown in Table 8.6. The design resistance is again represented by the inert reference strength $f_{0,\text{inert}}$ and the corresponding initial crack depth $a_i(f_{0,\text{inert}})$. In addition to what was seen from the figures, the table shows the following:

- The results ($f_{0,\text{inert}}$, $a_i(f_{0,\text{inert}})$) obtained from the various test series and design methods are extremely different.

- Because of the large influence of the scatter, many of the initial crack depths $a_i(f_{0,\text{inert}})$ are unrealistically high, especially for low $P_{f,t}$ and low m_0. This means that the mathematical model is prone to yielding results that are unrealistic from a physical point of view.

CHAPTER 8. STRUCTURAL DESIGN OF GLASS ELEMENTS

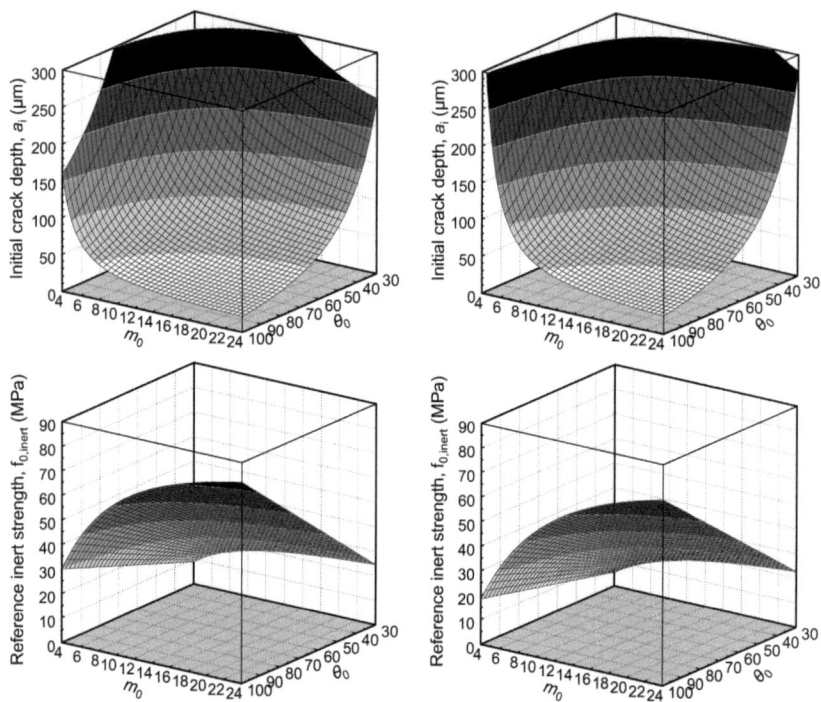

Figure 8.3: Initial crack depth a_i and reference inert strength $f_{0,\text{inert}}$ as a function of the surface condition parameters θ_0 and m_0. Left: $P_f = 0.008$, right: $P_f = 0.0012$.

- Artificially induced homogeneous surface damage causes the scatter and the mean of measured strength data to decrease (increasing m_0, decreasing θ_0). Since the influence of the scatter is predominant at low target failure probabilities, such damage actually increases design strength (reduces design crack depth) when compared to as-received glass.[1]

It now becomes clear why European and GFPM-based design methods adopt so different target failure probabilities. GFPM-based methods use a high target failure probability in combination with ambient strength data from weathered window glass specimens. European methods use the low target failure probability required by EN 1990:2002. Therefore, they cannot use strength data from weathered window glass, because this would yield an unrealistically low design resistance. To avoid this problem, ambient strength data from specimens with artificially induced homogeneous surface damage are used. Compared to the damage on weathered window glass, the homogeneous damage reduces the scatter of strength data markedly. As a consequence, m_0 becomes high enough to allow for the use of a low target failure probability without obtaining unrealistic results.

Figure 8.7 illustrates this issue. It can be seen that in terms of the reference inert strength or the corresponding initial crack depth, both approaches actually make very similar assumptions, with the US standard being only slightly more conservative.[2]

---

[1] *Example* (cf. Table 8.6): With $P_f = 0.0012$ (from EN 1990:2002), the design resistance of glass with artificially induced homogeneous surface damage is significantly higher than the design resistance of as-received glass according to the test series that DIN 1249-10:1990 is based on and far higher than the design resistance of weathered window glass according to ASTM E 1300-04. With $P_f = 0.008$, the 'extrapolation effect' is reduced. The resistance of homogeneously scratched glass is now comparable to that of as-received glass, but still higher than the resistance of weathered glass.

[2] This comparison is based on the reference area for $f_{0,\text{inert}}$, which is $1\,\text{m}^2$. Because of the different m_0-values used, the

8.3. CHARACTERISTICS OF THE TWO SURFACE CONDITION MODELS

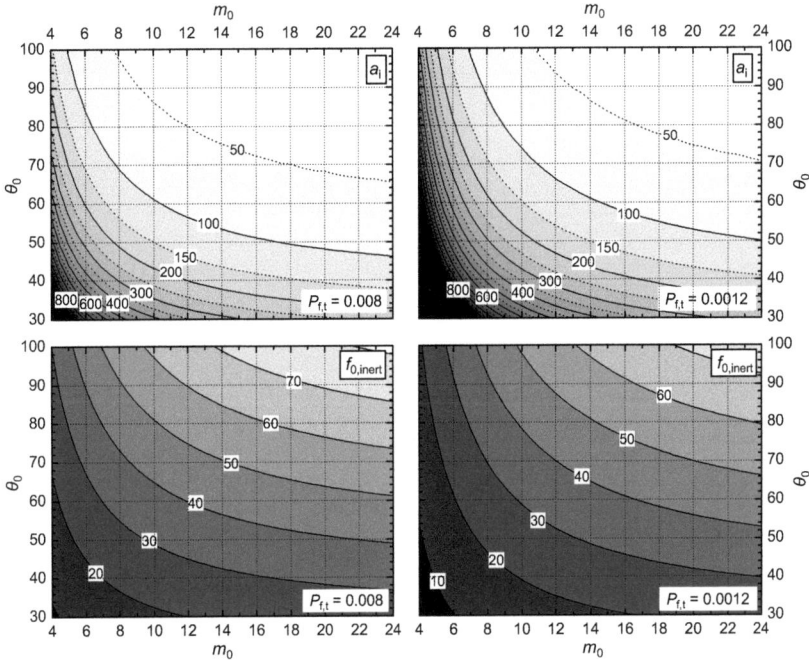

Figure 8.4: Initial crack depth a_i and reference inert strength $f_{0,\text{inert}}$ as a function of the surface condition parameters θ_0 and m_0. Left: $P_f = 0.008$, right: $P_f = 0.0012$ (Two-dimensional representation of Figure 8.3).

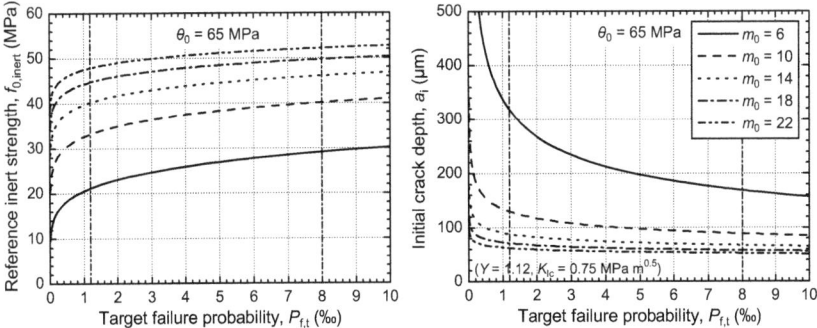

Figure 8.5: Reference inert strength $f_{0,\text{inert}}$ (left) and initial crack depth a_i (right) as a function of the target failure probability and the scatter of the surface condition data (m_0).

Table 8.6: Reference inert strength $f_{0,inert}$ and corresponding crack depth $a_i(f_{0,inert})$ for various test series and design methods.

	$\theta_0^\S$ (MPa)	m_0 (–)	$P_f = 0.008$		$P_f = 0.0012$	
			$f_{0,inert}$ (MPa)	$a_i(f_{0,inert})$ (μm)	$f_{0,inert}$ (MPa)	$a_i(f_{0,inert})$ (μm)
As-received glass						
DIN 1249-10:1990	62.89	4.94	23.69	254	16.13	549
Brown (1974)	70.21	6.39	32.99	131	24.50	238
Beason and Morgan (1984)	60.27	7.88	32.66	134	25.66	217
Fink (2000)	61.20	6.30	28.46	176	21.05	322
This study from ORF data[*†]	62.95	8.09	34.67	119	27.41	190
This study from inert tests[*‡]	67.57	7.19	34.54	120	26.52	203
Weathered window glass						
Beason (1980)	27.65	5.25	11.03	1173	7.68	2419
ASTM E 1300-04	40.93	6.13	18.62	412	13.65	766
Fink (2000)	20.82	3.76	5.78	4274	3.49	11734
Glass with artificially induced homogeneous surface damage						
DELR design method / prEN 13474	28.30	20.59	22.39	285	20.41	342
Blank (1993)	35.37	33.19	30.59	153	28.89	171
Blank (1993)	33.29	23.53	27.12	194	25.01	228

[*] Inert conditions, therefore independent of the crack velocity parameters
[†] See Section 6.5.
[‡] See Section 7.3.
[§] This surface condition parameter is derived from ambient strength data assuming $n = 16$ and $v_0 = 0.01$ mm/s. This is an approximate estimation of the crack velocity parameters during laboratory tests at medium stress rate, see Section 6.5.

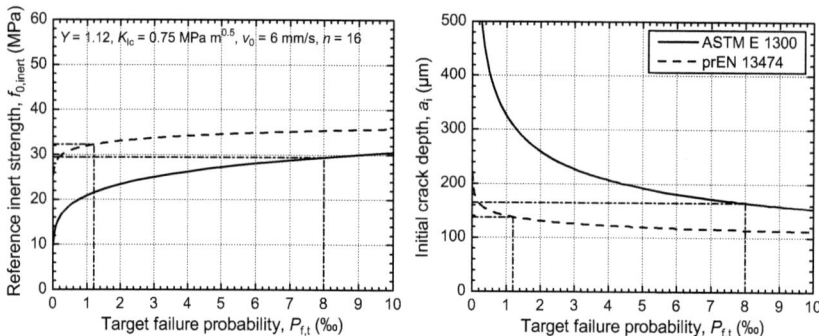

Figure 8.7: On the definition of the design resistance in European and GFPM-based standards.

Conclusions

In light of its derivation (Section 4.3), the assessment of the model hypotheses (Section 5.1) and the above considerations, the main characteristics of the random surface flaw population (RSFP) model can be summarized as follows:

- The model represents as-received glass and glass with artificially induced homogeneous surface damage well. It may, however, often be unrepresentative of in-service conditions, especially if deep surface flaws occur or if glass elements contain machining damage (cf. Section 5.1).

design resistance according to ASTM decreases faster than it does according to prEN for increasing surface area and vice versa.

- The model caters for arbitrary geometries and loading conditions, as long as the relevant crack opening stress history at all points on the surface can be determined.
- The approach accounts for the element's size and for biaxial and non-homogeneous stress fields.
- Surface damage hazard scenarios cannot easily be modelled.
- Numerical modelling is generally complex and computing time intensive. It can, however, be simplified significantly for many cases of practical relevance.
- RSFP-based modelling yields accurate results for likely values, i. e. at medium to high failure probabilities. The approach is, therefore, well-suited to the interpretation of laboratory tests. For structural design, it has a notable drawback. Because of the extrapolation to very low failure probabilities, the design resistance obtained is very sensitive to the scatter of the underlying strength data and the choice of the target failure probability. This inevitably leads glass designers to adjust the target failure probability and the model parameters in order to obtain results close to experience values rather than to design on a proper scientific basis.

8.4 Recommendations

8.4.1 Structural design

Structural design of glass elements with the lifetime prediction model generally involves the following steps:

1. Decide whether a single surface flaw (SSF) or a random surface flaw population (RSFP) is the more suitable surface condition model for the task at hand.
2. Determine the design crack depth $a_{i,d}$ (SSF) or the surface condition parameters θ_0 and m_0 (RSFP) by testing at inert conditions or using another suitable method (engineering judgement, detectability criteria, etc.).
3. Make conservative assumptions for the crack velocity parameters (n, v_0), the fracture mechanics parameters (K_{Ic}, Y) and the residual surface compression stress (σ_r).
4. Define a design action history and establish the action/stress relationship (normally by finite element analysis) for the location of the design flaw (SSF) or for all points on the element's surface (RSFP).
5. Assess the structural performance of the glass element and modify the design if required.

The following paragraphs give more specific recommendations for cases of practical relevance. It will be seen that only few cases actually require all of the above-mentioned steps to be carried out.

Exposed glass surfaces

- **Definition.** Exposed surfaces are glass surfaces that may be exposed to accidental impact, vandalism, heavy wind-borne debris or other factors that result in surface flaws that are substantially deeper than the 'natural' flaws caused by production and handling. Such flaws will be called 'severe damage' hereafter.
- **Surface condition model.** Structural design of glass elements with exposed surfaces should be based on a *design flaw*, which is a realistic estimation of the potential damage caused by surface damage hazards. Accordingly, the surface condition should be represented by a single surface flaw (cf. Section 8.3.1).
- **Long-term loading.** Long-term inherent strength in the presence of a deep design flaw is generally low (see left graph in Figure 8.2), has a large scatter and depends on many external influences (see Section 8.1). Therefore:

⋄ *Annealed glass* should not be relied upon for structural glass elements with long-term loads and exposed surfaces. If annealed glass must be used for some reason (cost, optical quality, tolerances, element size, etc.), failure consequences have to be evaluated carefully. Protection of building occupants in the case of glass breakage, post-breakage structural capacity, structural redundancy and easy accessibility for the replacement of broken glass elements become key aspects.

⋄ In the case of *heat strengthened or fully tempered glass*, the inherent strength in the presence of a design flaw is low compared to the residual stress, so that it contributes little to the effective resistance. In view of this limited structural benefit and the complex time-dependent behaviour, it is reasonable to ignore the inherent strength entirely and to design the glass element such that surface decompression is prevented at all points on the surface and during the entire service life (cf. Section 4.2.3). Because residual stresses do not depend on any external influences or on time, this kind of design is extremely simple.

- **Impact and short-term loading.** While neglecting the inherent strength is safe, it may in some cases be deemed too conservative for impact and short-term loads. In these cases, the inherent strength can be estimated as described above for annealed glass.
- **Quality control and inspections.** If information from inspections is to be used for less conservative design, periodic inspections during the entire service life are required. In the case of heat treated glass, quality control measures that make it possible to use a high design value for the residual surface stress are far more efficient in terms of economical material use than taking the inherent glass strength into account (with or without inspection).

Machining damage

Since the orientation and, more importantly, the location of the flaws are often not random (cf. Section 5.1.3), an RSFP-based model could produce unsafe results if glass elements contain significant machining damage. It is, therefore, recommended to design such elements using a design flaw that accounts for both machining damage and surface damage hazard scenarios.

Non-exposed glass surfaces

- **Definition.** *Non-exposed surfaces* are glass surfaces that are permanently and safely protected in all relevant hazard scenarios, so that they do not undergo any surface damage from external influences. Examples are inwardly oriented faces of laminated glass and insulating glass units, the surfaces of inner sheets of triple laminated glass and surfaces of glass elements that are installed in inaccessible locations.
- **Surface condition model.** RSFP-based modelling (cf. Section 8.3.2) is suitable for glass elements with non-exposed surfaces as long as loading conditions are rather simple and failure away from the edges is relevant. If edge failure is relevant or if complex loading conditions make the RSFP-based model too complex to be used with reasonable effort, SSF-based modelling is a conservative and much simpler alternative.
- **Quality control and inspections.** Inspections of the glass surface immediately after the installation of structural elements can provide useful information that allows for less conservative design. Although the inherent strength is much higher compared to that of exposed elements, quality control measures that make it possible to use a high design value for the residual surface stress of heat treated glass remain very efficient in terms of economical material use.

Non-structural glass elements

If failure and replacement of an element in the case of severe surface damage is accepted, non-structural elements can be designed as non-exposed structural elements. If non-structural elements have to withstand mainly lateral loads, design is, especially for heat strengthened or fully tempered glass, often governed by deflection criteria.

8.4.2 Testing

Despite the flexibility of the lifetime prediction model, it is unrealistic to expect that structural glass design can in all cases be solely based on predictive modelling. The difficulties with modelling arise mainly in the following areas:

- Glass is extremely sensitive to *stress concentrations*. Numerical models, however, often cannot provide reliable information on stress concentrations because it is difficult to model load introduction and support conditions with sufficient accuracy (substructure, liners, gaskets, bushings, contact stresses, etc.).
- Despite recent advances in the field (Kott, 2006), the *post-breakage structural capacity* often cannot be reliably predicted by predictive modelling.
- There is not much experience and quantitative information available concerning the *surface damage* caused by various hazard scenarios.
- The response of structural elements or entire sub-structures to *impact loads* is difficult to model.
- Building owners and authorities generally have *little confidence* in glass structures and often ask for full scale tests.

In conclusion, full-scale testing remains an integral part of the design process of innovative glass structures. The following should be considered:

- It is very important that design and interpretation of tests are based on a thorough understanding of the material behaviour. Especially the fact that results from tests at ambient conditions represent a combination of both surface condition and time-dependent crack growth is crucial. It is unfortunate that most project-specific testing is performed without taking time-dependent effects properly into account.
- If testing at ambient conditions is unavoidable, subcritical crack growth during the tests must be modelled. While this can efficiently be done using the model and the software tools presented in this thesis, the dependence of the crack velocity parameters on the environmental conditions and the stress rate still diminishes the accuracy and reliability of the results.
- The problems related to subcritical crack growth in laboratory tests can be addressed by the near-inert testing procedure developed in Chapter 7 and summarized below. By preventing crack growth during tests, it allows substantial improvement in the accuracy and safety of test results.
- Laboratory testing for structural elements that may be exposed to severe surface damage should be performed on specimens with realistic design flaws rather than on as-received or homogeneously damaged specimens. Suitable tools and testing procedures were developed in Section 7.4 and are summarized below. The key factor for meaningful results is a close match between the design flaw and potential in-service damage.

Determination of surface condition parameters

For random surface flaw population-based modelling and design, reliable surface condition parameters (θ_0, m_0) are required. Is was seen in Chapter 6 that this data should be obtained from laboratory testing in near-inert conditions in order to avoid the results being influenced by subcritical crack growth. A suitable testing procedure was developed and used in the present study, see Section 7.3.

The experimentally determined failure stresses represent the material's inert strength. The surface condition parameters can be obtained from such data as follows:

- For *tests with simple stress fields*, such as coaxial double ring tests or four point bending tests, the equations in Section 5.3.3 can be used. Fitting of the Weibull distribution to test results can be done by simple parameter estimation or maximum likelihood fitting (see Section E.3).

- For *tests with complex stress fields*, such as tests on large rectangular glass plates, the general lifetime prediction model should be used (see Section 6.5 for an example and further details, Section 8.5 for the software implementation of the model). Failure stresses are influenced by the non-linear load/stress relationship and the complex stress field. The data will therefore generally not follow a Weibull distribution and least-squares or maximum likelihood fitting (see Section E.3) should be used.

Assessment of the mechanical behaviour of design flaws

The testing procedure presented in Section 7.4 is suitable for obtaining experimental data for design flaw-based design. It can in particular be used to quantify the damage caused by a surface damage hazard scenario (design flaw) and to assess the structural performance of glass elements that contain such damage.

Figure 8.8 uses Equation (4.20) to show the expected failure stress in constant stress rate testing as a function of the flaw depth and the stress rate. It can be seen that in the case of slow crack velocity (right figure), the strength of deep surface flaws measured at ambient conditions and with a stress rate of 20 MPa/s or above is virtually identical to the inert strength. Since the crack velocity is indeed slow in common laboratory conditions (cf. Section 6.5), there should be no significant difference between specimens with and without hermetic coating. This was confirmed by experiments in Section 7.4.

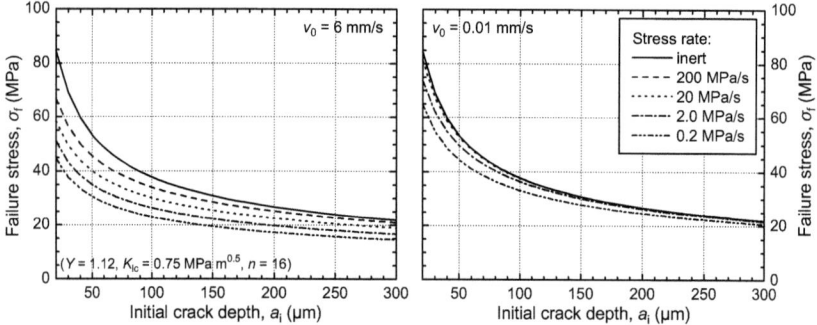

Figure 8.8: Failure stress in constant stress rate tests as a function of the initial crack depth and the stress rate for fast (left) and slow (right) crack velocity.

In conclusion, it is sufficient to ensure a stress rate in the order of magnitude of 20 MPa/s to obtain near-inert conditions in laboratory tests on specimens with deep surface flaws. Strength data measured in such tests can be interpreted as inert strength data without being excessively conservative. Since no surface drying and hermetic coating is required, laboratory testing for structural design based on surface damage hazard scenarios is simple and inexpensive, even in the case of large structural elements. The key factor for meaningful results is a close match between the design flaw and potential in-service damage.

8.4.3 Overview of mathematical relationships

An overview of the mathematical relationships to be used in the different cases is provided in Table 8.9.

8.4. RECOMMENDATIONS

Table 8.9: Table of mathematical relationships (SSF = single surface flaw, RSFP = random surface flaw population).

Structural design (simplified design equations)

SSF $\sigma_{t_0} \leq \sigma_{R,t_0}$

$$\sigma_{R,t_0} = \left(\frac{1}{t_0} \frac{2}{(n-2) \cdot v_0 \cdot K_{Ic}^{-n} \cdot (Y\sqrt{\pi})^n \cdot a_i^{(n-2)/2}} \right)^{1/n} \quad \rightarrow \text{Equation (4.23)}$$

$$\sigma_{t_0} = \left(\frac{1}{t_0} \int_0^T \sigma^n(\tau) \, d\tau \right)^{1/n} \approx \left(\frac{1}{t_0} \sum_{j=1}^J \left[\sigma_{t_j}^n \cdot t_j \right] \right)^{1/n} \quad \rightarrow \text{Equation (4.22)}$$

RSFP $\bar{\sigma} < f_0$

$$f_0 = \left[-\ln(1 - P_{f,t}) \right]^{1/\tilde{m}} \cdot \left(\frac{t_0}{U \cdot \theta_0^{n-2}} \right)^{-1/n} = f_{0,\text{inert}}^{(n-2)/n} \cdot \left(\frac{U}{t_0} \right)^{1/n} \quad \rightarrow \text{Equation (5.26)}$$

General cases:

$$\bar{\sigma} = \left(\frac{1}{A_0} \int_A \sigma_{1,t_0}^{\tilde{m}} \, dA \right)^{1/\tilde{m}} \approx \left(\frac{1}{A_0} \sum_i \left[A_i \cdot \sigma_{1,t_0,i}^{\tilde{m}} \right] \right)^{1/\tilde{m}} \quad \rightarrow \text{Equation (5.22)}$$

$$\sigma_{1,t_0,i} = \left(\frac{1}{t_0} \int_0^T \sigma_{1,i}^n \, d\tau \right)^{1/n} \approx \left(\frac{1}{t_0} \sum_{j=1}^J \left[\sigma_{1,\tau_j,i}^n \cdot \tau_j \right] \right)^{1/n} \quad \rightarrow \text{Equation (5.19)}$$

Simplification if the decompressed surface area remains constant and the major principal stress is proportional to the load at all points on the surface and during the entire loading history:

$$\bar{\sigma} = A_0^{-1/\tilde{m}} \cdot \mathbb{X}_{t_0} \cdot \bar{A}^{1/\tilde{m}} \quad \text{with} \quad \bar{A} = \int_A c(\vec{r})^{\tilde{m}} \, dA \approx \sum_i A_i c_i^{\tilde{m}} \quad \rightarrow \text{Equation (5.31)}$$

$$\mathbb{X}_{t_0} = \left(\frac{1}{t_0} \int_0^T \mathbb{X}^n(\tau) \, d\tau \right)^{1/n} \approx \left(\frac{1}{t_0} \sum_{j=1}^J \left[\mathbb{X}_{t_j}^n \cdot t_j \right] \right)^{1/n} \quad \text{with} \quad \mathbb{X} = \begin{cases} \bar{\sigma} & \text{for stresses} \\ q & \text{for loads} \end{cases} \quad \rightarrow \text{Equation (5.35)}$$

Interpretation of tests (general equations for predictive modelling)

SSF $\tilde{a}_c$

$$\tilde{a}_c(\tau) = \left[\left[\frac{(\sigma_n(\tau) \cdot Y \sqrt{\pi}}{K_{Ic}} \right]^{n-2} + \frac{n-2}{2} v_0 \cdot K_{Ic}^{-n} \cdot (Y\sqrt{\pi})^n \cdot \int_0^\tau \sigma_n^n(\tilde{\tau}) \, d\tilde{\tau} \right]^{\frac{2}{2-n}} \quad \rightarrow \text{Equation (4.20)}$$

RSFP P_f

General cases:

$$P_f(t) = 1 - \exp\left\{ -\frac{1}{A_0} \int_A \frac{2}{\pi} \int_{\varphi=0}^{\pi/2} \left[\max_{\tau \in [0,t]} \left[\left(\frac{\sigma_n(\tau,\vec{r},\varphi)}{\theta_0} \right)^{n-2} + \frac{1}{U \cdot \theta_0^{n-2}} \int_0^\tau \sigma_n^n(\tilde{\tau},\vec{r},\varphi) \, d\tilde{\tau} \right]^{\frac{1}{n-2}} \right]^{m_0} \, dA \, d\varphi \right\} \quad \rightarrow \text{Equation (4.77)}$$

For simplifications for common test setups with uniform stress fields and for testing at inert conditions, see Section 5.3.3.

8.5 Software

8.5.1 Introduction and scope

Until recently, the only way of enabling complex engineering models to be used in practice was to provide tables, graphs, simplified equations or rules of thumb. Nowadays, there is another possibility. Software can handle complex engineering models in a more convenient, fast and straightforward way than the aforementioned approaches without adversely affecting accuracy or flexibility.

The computer software, called *GlassTools*, which was developed as part of this thesis, aims to provide convenient and flexible tools to perform calculations and simulations using the models and concepts discussed in this document. The main design objectives for the software are as follows:

1. *GlassTools* should be able to be used for deterministic and probabilistic calculations.

2. It should be based on modern, widely used software development technology.

3. No commercial software should be required to run *GlassTools*.

4. *GlassTools* should target the Microsoft Windows platform and integrate seamlessly with Microsoft Excel (MS Excel 2003), because these tools are predominantly used in the field of structural engineering.

5. Since Windows and Excel are commercial applications, GlassTools' core functionality must be available without Excel and on non-Windows operating systems in order to meet objective 3.

6. *GlassTools* should provide good performance on current mainstream hardware.

7. It should be suitable as a basis for customized applications and future developments. This requires in particular a clean design and comprehensive code documentation.

Since *GlassTools* is research software developed with very limited resources, it does *not* aim at providing an elaborate graphical user interface, convenience functions, features for result visualization, a high level of robustness or extensive error handling.

8.5.2 Discretization of the failure prediction model

The numerical implementation of the failure prediction model requires its discretization. While this is a straightforward process for some aspects of the model, others deserve further consideration. This section focuses on these aspects.

Random surface flaw population-based modelling

Discretization of Equation (4.77) can be obtained as follows:

- The total surface area A of the glass element is divided into I sub-elements of area A_i ($i = 1, 2, \ldots, I$; $\sum A_i = A$) and centroid coordinates (x_i, y_i). The sub-elements must be so small that the stress field is approximately homogeneous within each sub-element.

- The survival probability P_s of the glass element is the product of the survival probabilities of all sub-elements: $P_s = \prod_{i=1}^{I} P_{s,i}$.

- The action history $q(\tau)$ with $\tau \in [0, t]$ is divided into J time intervals of duration τ_j ($j = 1, 2, \ldots, J$; $\sum \tau_j = t$). The time intervals must be so short that the action intensity is approximately constant during each interval.

- In order to account for crack orientation, the quarter circle is represented by K segments of central angle $\Delta\varphi_k$ each ($k = 1, 2, \ldots, K$; $\sum \Delta\varphi_k = \pi/2$). Crack orientation is defined as follows: A crack with orientation φ_k lies in a plane perpendicular to the bisectrix of the segment $\Delta\varphi_k$. The segments must be so small that all cracks of orientation $\varphi_k - \Delta\varphi_k/2 \leq \varphi_{\text{crack}} < \varphi_k + \Delta\varphi_k/2$ are exposed to approximately the same crack opening stress $\sigma_n(\varphi_k)$.

8.5. SOFTWARE

Consequently, all cracks on the surface of the i-th sub-element with orientation $\varphi_k - \Delta\varphi_k/2 \leq \varphi_{\text{crack}} < \varphi_k + \Delta\varphi_k/2$ are exposed to the constant crack-opening stress $\sigma_n(i,j,k)$ during the time interval τ_j. The failure probability using this discrete formulation is:

$$P_f(t) = 1 - \exp\left\{-\frac{1}{A_0}\sum_{i=1}^{I}\frac{2}{\pi}\sum_{k=1}^{K}\left[\max_j\left\{\left(\frac{\sigma_n(i,j,k)}{\theta_0}\right)^{n-2} + \frac{1}{U\cdot\theta_0^{n-2}}\sum_{j=1}^{J}\left(\sigma_n(i,j,k)^n\cdot\tau_j\right)\right\}^{\frac{1}{n-2}}\right]^{m_0}\Delta\varphi_k\, A_i\right\} \quad (8.2)$$

The model parameters were already explained below Equation (4.77). In addition to these, the quantification of Equation (8.2) requires:

- The results of a transient, generally geometrically non-linear *finite element calculation*. These results must provide the surface areas A_i of all sub-elements and the major and minor principal stresses $\sigma_1(i,q_j)$ and $\sigma_2(i,q_j)$ for all sub-elements i and at all action intensities q_j.
- A probabilistic or deterministic *design load model* for the design service life of the structural component.

Inert conditions

The inert failure probability is obtained by removing the subcritical crack growth term from Equation (8.2):

$$P_{f,\text{inert}}(t) = 1 - \exp\left\{-\frac{1}{A_0}\sum_{i=1}^{I}\frac{2}{\pi}\sum_{k=1}^{K}\left[\left(\frac{\max_j\{\sigma_n(i,j,k)\}}{\theta_0}\right)^{m_0}\Delta\varphi_k\, A_i\right]\right\} \quad (8.3)$$

As expected, Equation (8.2) converges on this result for $t = \sum \tau_j \to 0\,\text{s}$, $v_0 \to 0\,\text{m/s}$ or $n \to \infty$. No specific implementation is, therefore, needed for inert conditions.

Approximation by considering damage accumulation only

It was seen in Section 5.2.1 that for sufficiently long loading times and sufficiently high values of v_0, Equation (4.77) can be simplified by considering the damage accumulation term only. With this approach, Equation (8.2) simplifies to:[3]

$$P_{f,\text{SCGonly}}(t) = 1 - \exp\left\{-\frac{1}{A_0}\sum_{i=1}^{I}\frac{2}{\pi}\sum_{k=1}^{K}\left[\left(\frac{1}{U\cdot\theta_0^{n-2}}\sum_{j=1}^{J}\left(\sigma_n(i,j,k)^n\cdot\tau_j\right)\right)^{\frac{m_0}{n-2}}\Delta\varphi_k\, A_i\right]\right\} \quad (8.4)$$

Biaxial stress fields

It was found in Section 5.2.3 that for most *design tasks*, an equibiaxial stress field may be assumed. In such a stress field, the two principal stresses are equal ($\sigma_1 = \sigma_2$) and the crack opening stress σ_n is orientation-independent. No segmenting is therefore required and $\Delta\varphi_k$ is simply $\pi/2$. In addition to the substantial reduction in computing time, this has the advantage that no σ_2 input data are required. When interpreting *experimental data*, however, the biaxial stress field must be taken into account (cf. Section 5.2.3). In order to determine the required number of segments K, Figure 8.10 shows the predicted failure probability for the test setup used in Section 6.5 with a pressure rate

---

[3] Equation (8.4) is given for clarity. It is, however, inefficient because several time-consuming mathematical operations on constants are performed repeatedly in the loops required to calculate the sums. The actual implementation in *GlassTools* avoids this.

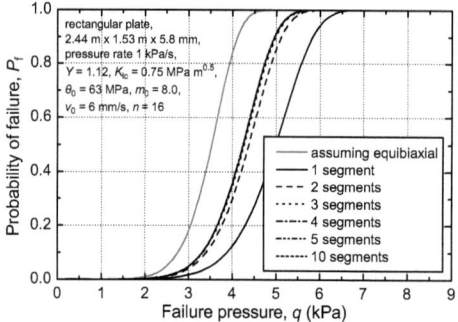

Figure 8.10:
Failure probability as a function of the number of segments in a quarter circle.

of 1 kPa/s. The figure shows that if an equibiaxial stress field is assumed, the resistance is clearly underestimated. It can also be seen that the results converge rapidly as the number of segments K increases. Five segments is a sensible choice.

Equivalent static stress

Some convergence issues with respect to the numerical determination of the equivalent t_0-second stress need to be clarified. If a stress history consists of J time intervals of duration τ_j and with constant stress σ_j, which is the output of most probabilistic load models, the formulation

$$\sigma_{t_0}(t) = \left(\frac{1}{t_0} \int_0^t \sigma^n(\tau) \, d\tau \right)^{1/n} \approx \left(\frac{1}{t_0} \sum_{j=1}^{J} \left[\sigma_j^n \cdot \tau_j \right] \right)^{1/n} \tag{8.5}$$

(cf. Equation (4.22)) is accurate. In laboratory testing however, there will generally be a constant displacement rate, a constant load rate or a constant stress rate. All three result in a rather smooth $\sigma(t)$ curve that is at least piecewise very close to the constant stress rate case ($\sigma(t) = \dot{\sigma} \cdot t$). It would be convenient if Equation (8.5), rather than some specialized formulation, could be used for these cases. To achieve an optimal trade-off between computing time and accuracy, the number of intervals J required to make the discrete formulation (right part of Equation (8.5)) converge on the exact one (left part) needs to be determined. If $\sigma(t) = \dot{\sigma} \cdot t + \sigma_0$ ($\sigma_0 = \sigma(t=0)$), the exact solution is:

$$\sigma_{\text{exact},t_0}(t) = \left(\frac{1}{t_0} \int_0^t (\dot{\sigma} \cdot \tau + \sigma_0)^n \, d\tau \right)^{1/n} = \left(\frac{1}{t_0} \frac{(\dot{\sigma} \cdot t + \sigma_0)^{n+1} - \sigma_0^{n+1}}{\dot{\sigma}(n+1)} \right)^{1/n} \tag{8.6}$$

Figure 8.11 shows the relation $\sigma_{t_0}/\sigma_{\text{exact},t_0}$ as a function of the number of intervals J for the most common case of $\sigma_0 = 0$. The result is independent of $\dot{\sigma}$ and t but depends slightly on the exponential crack velocity parameter n. Curves for several n values are therefore plotted.

8.5.3 Implementation

Software development is beyond the scope of this thesis. With this in mind, the present section is by no means exhaustive, but sheds light on some key aspects and concepts that may be of interest to the reader. Several concepts and technologies will be mentioned without detailed explanation. The interested reader can find exhaustive information on the internet. The free online encyclopedia Wikipedia, for instance, is an ideal starting point.

The design objectives outlined in Section 8.5.1 already restrict the choice of programming languages to only a few options. One more key aspect is the performance of computing time intensive numeric calculations. Programming languages benchmarking is generally a complex task and results depend

8.5. SOFTWARE

Figure 8.11:
Convergence of the numeric equivalent static stress calculation for a constant stress rate.

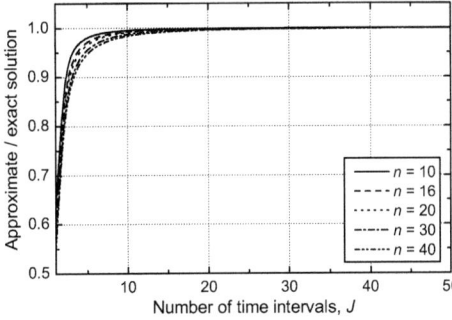

Figure 8.12:
Performance comparison of C#, C++, Matlab, and Python.

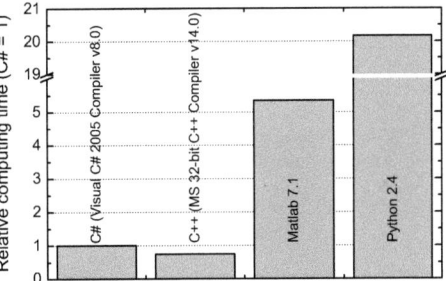

strongly on the test setup and criteria. For the specific task at hand, however, it is sufficient to use a simple but focused approach that consists of measuring the computing time of a small algorithm that performs the main operations required to use the lifetime prediction model. The performance of this algorithm reflects the performance to be expected when running the lifetime prediction model. The benchmarking algorithm involves the following steps[4]: (1) allocate two arrays with 10^5 double precision floating point elements each; (2) fill the first array using integer to double type casts, multiplication and division; (3) fill the second array with the values from the first raised to a power; (4) add up all elements of the second array.

Figure 8.12 shows the results of the performance comparison. The following can be seen:

- As expected, the **C++** implementation performs best.

- Interestingly, however, **C#** is only marginally slower. This means that C#, which runs as managed code on top of the .NET Common Language Runtime (CLR) and involves just-in-time compilation, is almost as suitable for these kinds of numerical calculations as the compiled-to-machine code language C++.

- The interpreted language **Python** is convenient and widely used for prototyping scientific algorithms. By virtue of the built-in high-level functionality and the availability of advanced open source libraries, development in Python is particularly simple and rapid. The benchmarking algorithm, however, requires twenty times more computing time in Python than in C#. This is a very severe drawback for the task at hand. Furthermore, providing Excel integration (cf. Section 8.5.1) with Python would be difficult.

- **Matlab**, an expensive commercial software package[5] (Matlab 2005), is both an application and a programming language. Since the design objectives for *GlassTools* require freely available technology, Matlab is not an option. It is nevertheless included in this comparison because it is used in Section 6.5.

---
[4] As an example, the C++ version of the algorithm is reproduced in Section G.2.
[5] There is a free *partial* clone of Matlab called GNU Octave.

On the basis of these results, it was decided to implement *GlassTools* in C#[6]. The significant advantages of this modern programming language clearly outweigh the marginally inferior performance when compared to C++.

8.5.4 Features and usage notes

In the following, a very short overview of *GlassTools*' main components and their functionality is given. For detailed documentation, the reader should refer to *GlassTools*' documentation in Appendix H.

- **GlassFunctionsForExcel** is an 'automation add-in' for Microsoft Excel that contains dozens of functions. They can be used exactly as the built-in functions in Excel worksheets and implement a subset of GlassTools' functionalities as well as most of the equations used in the present document. Functions that require finite element data (action/stress relationship and surface area for all finite elements) read them from text files. The file paths are supplied as arguments.

  In addition to being used in Excel worksheets, the functions can also be accessed by any Windows program or programming language through COM[7]. A Windows installer for GlassFunctionsForExcel is provided, enabling installation with just a few mouse clicks.

- A **set of classes** provide the functionality required to perform calculations and simulations using the model and related concepts discussed in this thesis. The main classes are:

  ⋄ **SSFM and RSFPM:** These classes represent the failure prediction model, using a single surface flaw (SSFM = single surface flaw model) and a random surface flaw population (RSFPM = random surface flaw population model) respectively to represent the surface condition. Their methods perform most calculations and simulations mentioned in the present document. Model input and parts of the functionality are made available by instances[8] of ActionObject, MaterialObject and GlassElement (see below).

  ⋄ **ActionObject:** This class represents a probabilistic or deterministic action model and generates action histories from this model. Random variables are represented by instances of the probability distribution classes.

  ⋄ **MaterialObject:** This class represents a probabilistic or deterministic material model. Random variables are represented by instances of the probability distribution classes.

  ⋄ **GlassElement:** This class represents a glass element. Its instances normally contain the finite element data representing the action/stress relationship and the surface area of all finite elements. The main purpose of the class is to calculate stresses and equivalent stresses as a function of an action or action history.

---

[6] Despite the fact that Microsoft is the driving force behind this technology, C# and its virtual machine* are not proprietary technologies but international standards (ISO/IEC 23270:2003; ISO/IEC 23271:2003). For development on the Windows platform, the .NET Framework SDK[†] is required. It is not open source, but can be downloaded free of charge (.NET Framework website). It includes, among others, documentation and the C# compiler. A basic but sufficient integrated development environment is also available free of charge (VS Express website). The Mono Project (www.mono-project.com) provides open-source software to develop, compile and run C# applications on Windows, Linux, Solaris, Mac OS X and other Unix-like operating systems.

[*] In general terms, a virtual machine is software that creates a virtualized environment between the computer platform and the end user in which the end user can operate software. The virtual machine at the heart of the Microsoft .NET initiative, called Common Language Infrastructure, emulates hardware that only exists as a detailed specification. This technique allows diverse computers to run any software written to that specification. Only the virtual machine software itself must be written separately for each type of computer on which it runs.

[†] A set of development tools (SDK = software development kit) that allows the creation of applications for the .NET framework using the C#, C++, VisualBasic or J# programming languages.

[7] COM (Component Object Model) is a Microsoft software platform. It is used to enable interprocess communication and dynamic object creation in any programming language that supports the technology.

[8] In object-oriented programming, an object that is created from a class is called an instance of that class.

⋄ **ProbabilityDistribution** and its derived classes[9]: These classes represent normal, lognormal, uniform and Weibull distribution functions. They are intended to be used as input parameters in probabilistic calculations. Their member functions include the generation of random numbers that follow the respective distribution type as well as the calculation of CDF, PDF, mean, variance etc.
* A **Windows application** with a graphical user interface. It aims to provide an example of the use of the *GlassTools* class library. Furthermore, it enables the non-programmer to perform some calculations that cannot be done in Excel without having to write program code.

An example that demonstrates the application of *GlassTools* for the prediction of the results of column buckling laboratory tests is provided in Appendix F.

8.6 Action modelling

Models for the analysis of structural components generally require a resistance model and an action model. The particularity of glass is that the inherent strength depends on subcritical crack growth and therefore on the entire stress history. Consequently, glass design requires action models that represent the entire action history if the inherent strength is to be taken into account. This is a fundamental difference from other materials such as steel or concrete, for which only extreme values or extreme value distributions are normally used for design (except for fatigue loads). On the other hand, subcritical crack growth depends on the stress raised to the exponent $n \approx 16$ (cf. Equations (4.22) and (4.77)), which is why the behaviour of the action model at its upper end remains of vital importance.

However, if the inherent strength is neglected as suggested above for the majority of cases, the issue is of no interest because extreme value models are sufficient for design based on avoiding surface decompression.

In conclusion, advanced action modelling for glass is, on one hand, an interesting and complex topic that clearly requires further research. On the other hand, its use is confined to relatively few cases and the inevitably complex models are unlikely to be adopted by engineers and the construction industry.

---

[9] In object-oriented programming, inheritance is a way to form new classes using classes that have already been defined. The former, known as derived classes, take over (or inherit) attributes and behaviour of the latter, which are referred to as base classes.

8.7 Summary and Conclusions

⇨ **Key aspects of the structural behaviour of glass elements.** The following key aspects govern the structural behaviour of glass elements and cause their structural design to differ considerably from the design of elements made of more common construction materials:

- **Surface condition.** The resistance of a glass element is very sensitive to the flaws on its surface. In addition to standard hazard scenarios, surface damage hazard scenarios should, therefore, be considered for design.
- **Subcritical crack growth.** Stress corrosion causes flaws on a glass surface to grow subcritically. The resistance of a loaded glass element therefore decreases with time, even if it is exposed only to static loads. The growth of a surface flaw depends on the properties of the flaw and the glass (initial crack depth, geometry factor, crack growth threshold), the crack opening stress history that the flaw is exposed to (which is a function of the element's geometry, the support conditions and the loading), and on the relationship between crack velocity and stress intensity (represented by the crack velocity parameters). The strong dependence of the crack velocity parameters on external influences such as temperature, humidity and the loading rate make accurate predictive modelling difficult. For structural design, a conservative estimate of the crack velocity parameters can be used. Experiments should be performed at inert conditions in order to prevent subcritical crack growth and its barely predictable effect on the results.
- **Surface decompression.** In surface regions that are not decompressed, there is no positive crack opening stress and consequently no subcritical crack growth. This is of particular importance for heat treated glass, in which surface decompression only occurs if the tensile stress due to loading exceeds the residual surface compression stress.

⇨ **Using the lifetime prediction model for structural design.** Structural design with the lifetime prediction model generally involves the following steps:

- Decide whether a single surface flaw (SSF) or a random surface flaw population (RSFP) is the more suitable surface condition model for the task at hand.
- Determine the design crack depth $a_{i,d}$ (SSF) or the surface condition parameters θ_0 and m_0 (RSFP) by testing at inert conditions or using another suitable method (engineering judgement, detectability criteria, etc.).
- Make conservative assumptions for the crack velocity parameters (n, v_0), the fracture mechanics parameters (K_{Ic}, Y) and the residual surface compression stress (σ_r).
- Define a design action history and establish the action/stress relationship (normally by finite element analysis) for the location of the design flaw (SSF) or for all points on the element's surface (RSFP).
- Assess the structural performance of the glass element and modify the design if required.

⇨ **Characteristics of the single surface flaw (SSF) model.** This model is suitable for cases in which a glass element's failure is governed by one deep surface flaw or a few such flaws. It caters for arbitrary geometries and loading conditions, as long as sensible assumptions with regard to the location of the design crack and the crack opening stress history at this location can be made. Since the outcome of the model is a function of the conditions at the location of the design flaw(s) only, it is not influenced by the element's size or by biaxial or non-homogeneous stress fields. Because of the simple representation of the surface condition, the model is intuitive, easy to use and numerical modelling is simple and fast. Surface damage hazard scenarios can be modelled with ease and information from testing, inspection, proof loading or engineering judgement can easily be integrated into the model.

8.7. SUMMARY AND CONCLUSIONS

⇨ **Characteristics of the random surface flaw population (RSFP) model.** The model represents as-received glass and glass with artificially induced homogeneous surface damage well. It may, however, often be unrepresentative of in-service conditions, especially if deep surface flaws occur or if glass elements contain machining damage. The model caters for arbitrary geometries and loading conditions, as long as the relevant crack opening stress history at all points on the surface can be determined. The approach accounts for the element's size and for biaxial and non-homogeneous stress fields. Numerical modelling is generally complex and computing time intensive. RSFP-based modelling yields accurate results for likely values, i.e. at medium to high failure probabilities. The approach is, therefore, well-suited to the interpretation of laboratory tests. For structural design, it has a notable drawback. Because of the extrapolation to very low failure probabilities, the design resistance obtained is very sensitive to the scatter of the underlying strength data and the choice of the target failure probability.

⇨ **Recommendations for the design of exposed glass surfaces.** Exposed surfaces are glass surfaces that may be exposed to accidental impact, vandalism, heavy wind-borne debris or other factors that result in surface flaws that are substantially deeper than the 'natural' flaws caused by production and handling. Structural design of glass elements with such surfaces should be based on a design flaw which is a realistic estimation of the potential damage caused by surface damage hazards. The long-term inherent strength in the presence of a deep design flaw is generally low, has a large scatter and depends on many external influences. *Annealed glass* should, therefore, not be relied upon for structural glass elements with long-term loads and exposed surfaces. If annealed glass must be used for some reason (cost, optical quality, tolerances, element size, etc.), failure consequences have to be evaluated carefully. Protection of building occupants in the case of glass breakage, post-breakage structural capacity, structural redundancy and easy accessibility for the replacement of broken glass elements become key aspects. In the case of *heat strengthened or fully tempered glass*, the inherent strength contributes little to the effective resistance. In view of this limited structural benefit and the complex time-dependent behaviour, it is in most cases reasonable to ignore it entirely and to design the glass element such that surface decompression is prevented. Because residual stresses do not depend on any external influences or on time, this kind of design is extremely simple.

⇨ **Recommendations in the case of machining damage.** Since the orientation and, more importantly, the location of the flaws are often not random, an RSFP-based model could produce unsafe results if glass elements contain significant machining damage. It is, therefore, recommended to design such elements using a design flaw that accounts for both machining damage and surface damage hazard scenarios.

⇨ **Recommendations for the design of non-exposed glass surfaces.** If a glass element's surfaces are permanently and safely protected from damage, and failure away from the edges is relevant, an RSFP-based model is suitable. *GlassTools*, the computer software that was developed as part of the present research, simplifies lifetime prediction and structural design with this model considerably. Nevertheless, the task may become difficult and time-consuming under complex loading conditions. In these cases, design flaw-based modelling is a useful, conservative and much simpler alternative.

⇨ **Quality control.** In the case of heat treated glass, quality control measures that make it possible to use a high design value for the residual surface stress are very efficient in terms of economical material use. For exposed glass elements, the potential gain in design strength is far higher than what can be gained by taking the inherent glass strength into account. And even in the case of protected elements, the gain is very substantial.

⇨ **Recommendations for testing.** Full-scale testing remains an integral part of the design process of innovative glass structures. The main reasons for this are the material's extreme sensitivity to stress concentrations, the difficulty in assessing the post-breakage performance and the response to impact loads using predictive modelling, the insufficient information about the surface damage caused by hazards and the fact that building owners and authorities generally have little confidence in glass structures and therefore ask for full scale tests. The following should be considered:

- It is very important that design and interpretation of tests be based on a thorough understanding of the material behaviour. The fact that results from tests at ambient conditions represent a combination of both surface condition and time-dependent crack growth is particularly crucial.
- If testing at ambient conditions is unavoidable, subcritical crack growth during the tests must be modelled. While this can efficiently be done using the model and the software tools presented in this thesis, the dependence of the crack velocity parameters on the environmental conditions and the stress rate still diminishes the accuracy and reliability of the results. The problem can be addressed by the near-inert testing procedure that was developed and used in this thesis. By preventing crack growth during tests, it allows substantial improvement in the accuracy and safety of test results.
- Laboratory testing for structural elements that may be exposed to severe surface damage should be performed on specimens with realistic design flaws rather than on as-received or homogeneously damaged specimens. Suitable tools and testing procedures were developed in this thesis.

⇨ **Computer software.** The lifetime prediction model is too complex for manual calculations. Computer software, called *GlassTools*, was therefore developed as part of the present research. It allows calculations and simulations based on the models and concepts discussed in this thesis to be performed conveniently and quickly. GlassTools consists of a C# class library with a COM-interface. Furthermore, a large part of the functionality is provided as an 'automation add-in' and can thus be used in Microsoft Excel. Comprehensive source code documentation is provided.

Chapter

9

Summary, Conclusions, and Further Research

This chapter summarizes the principal findings and gives the most significant conclusions from the work. Finally, recommendations for future work are made.

9.1 Summary

A short overview of the content of this thesis is provided in the abstract on page i and in Section 1.3 on page 3. The main findings are summarized in the grey boxes at the end of each chapter, namely on pages 35, 56, 78, 95, 114 and 134. The interested reader should refer to these sections for an overview of the work.

9.2 How this work's objectives were reached

Below, it is briefly discussed how the four objectives that were given in Section 1.2 were reached.

Objective 1: To analyse the current knowledge about, and identify research needs for, the safe and economical structural design of glass elements.

A comprehensive analysis of present knowledge was conducted and provided a focus for subsequent investigations. The bases, underlying hypotheses, advantages, limits and shortcomings of current glass design methods were presented and discussed in detail.

Objective 2: To establish a lifetime prediction model for structural glass elements, which is as consistent and flexible as possible and offers a wide field of application. [...]

Such a model was established based on fracture mechanics and the theory of probability. It offers significant advantages over currently used models.

Objective 3: To assess the main hypotheses of the model, discuss possibilities for its simplification and relate it to existing models. To verify the availability of the required model input data and improve existing testing procedures in order to provide more reliable and accurate model input.

Possible simplifications of the model, its relation to existing models and the availability of the model's required input data were discussed. In addition to the analysis of existing data, laboratory tests were performed and testing procedures improved in order to provide more reliable and accurate model input.

Objective 4: To deploy the model to provide recommendations for the structural design of glass elements and for the laboratory testing required within the design process. To develop computer software that facilitates the application of the recommendations and enables the model to be used efficiently in research and practice.

The findings of the present research allowed such recommendations to be made and computer software that meets this objective to be developed.

9.3 Main conclusions

The overall objective of this thesis was to investigate how structural glass elements of arbitrary geometry can be conveniently and accurately analysed and designed. The key findings with respect to this aim are summarized in the following section.

1. **Key aspects of the structural behaviour of glass elements.** The following key aspects govern the structural behaviour of glass elements and cause their structural design to differ considerably from the design of elements made of more common construction materials:

   - **Surface condition.** The resistance of a glass element is very sensitive to the flaws on its surface. In addition to standard hazard scenarios, surface damage hazard scenarios should, therefore, be considered for design.

   - **Subcritical crack growth.** Stress corrosion causes flaws on a glass surface to grow subcritically. The resistance of a loaded glass element therefore decreases with time, even if it is exposed only to static loads. The growth of a surface flaw depends on the properties of the flaw and the glass (initial crack depth, geometry factor, crack growth threshold), the crack opening stress history that the flaw is exposed to (which is a function of the element's geometry, the support conditions and the loading), and on the relationship between crack velocity and stress intensity (represented by the crack velocity parameters). The heavy dependence of the crack velocity parameters on external influences such as temperature, humidity and the loading rate make accurate predictive modelling difficult. For structural design, a conservative estimate of the crack velocity parameters can be used. Experiments should be performed at inert conditions in order to prevent subcritical crack growth and its barely predictable effect on the results.

   - **Surface decompression.** In surface regions that are not decompressed, there is no positive crack opening stress and consequently no subcritical crack growth. This is of particular importance for heat treated glass, in which surface decompression only occurs if the tensile stress due to loading exceeds the residual surface compression stress.

2. **Lifetime prediction model.** A lifetime prediction model for structural glass elements was derived based on fracture mechanics and the theory of probability. This model can be used for predictive modelling and for structural design. Aiming at consistency, flexibility, and a wide field of application, it offers significant advantages over currently used models:

   - The model takes all aspects that influence the lifetime of structural glass elements into consideration. It contains no simplifying hypotheses which would restrict its applicability to special cases. It is, therefore, valid for structural glass elements of arbitrary geometry and loading, including in-plane or concentrated loads, loads which cause not only time-variant stress levels but also time-variant stress distributions, stability problems, and connections.

   - The parameters of the model have a clear physical meaning that is apparent to the engineer. They each include only one physical aspect and they do not depend on the experimental setup used for their determination.

   - The condition of the glass surface can be modelled using either a single surface flaw (SSF) or a random surface flaw population (RSFP), i.e. a large number of flaws of random depth, location and orientation. The properties of these surface condition models are independent parameters that the user can modify. This is a major advantage, especially when hazard scenarios that involve surface damage must be analysed or when data from quality control measures or research are available.

9.3. MAIN CONCLUSIONS

- The model does not suffer from the two problems of current glass design methods that gave rise to fundamental doubts regarding advanced glass modelling: The material strength rightly converges on the inert strength for very short loading times or slow crack velocity and the momentary failure probability is not independent of the momentary load.

3. **Representation of a glass element's surface condition.** As mentioned above, the lifetime prediction model offers two different surface condition models:
   - The *single surface flaw (SSF) model* is suitable for cases in which a glass element's failure is governed by one deep surface flaw or a few such flaws. Because of the simple representation of the surface condition, the model is intuitive and easy to use. Surface damage hazard scenarios can be modelled with ease and information from testing, inspection, proof loading or engineering judgement can easily be integrated into the model.
   - The *random surface flaw population (RSFP) model* represents as-received glass and glass with artificially induced homogeneous surface damage well. It may, however, often be unrepresentative of in-service conditions, especially if deep surface flaws occur or if glass elements contain machining damage. The approach accounts for the element's size and for biaxial and non-homogeneous stress fields. The approach is well-suited to the interpretation of laboratory tests. For structural design, it has a notable drawback. Because of the extrapolation to very low failure probabilities, the design resistance obtained is very sensitive to the scatter of the underlying strength data and the choice of the target failure probability.

4. **Recommendations for the design of exposed glass surfaces.** Exposed surfaces are glass surfaces that may be exposed to accidental impact, vandalism, heavy wind-borne debris or other factors that result in surface flaws that are substantially deeper than the 'natural' flaws caused by production and handling. Structural design of glass elements with such surfaces should be based on a design flaw which is a realistic estimation of the potential damage caused by surface damage hazards. The long-term inherent strength in the presence of a deep design flaw is generally low, has a large scatter and depends on many external influences. *Annealed glass* should, therefore, not be relied upon for structural glass elements with long-term loads and exposed surfaces. If annealed glass must be used for some reason (cost, optical quality, tolerances, element size, etc.), failure consequences have to be evaluated carefully. Protection of building occupants in the case of glass breakage, post-breakage structural capacity, structural redundancy and easy accessibility for the replacement of broken glass elements become key aspects. In the case of *heat strengthened or fully tempered glass*, the inherent strength contributes little to the effective resistance. In view of this limited structural benefit and the complex time-dependent behaviour, it is in most cases reasonable to ignore it entirely and to design the glass element such that surface decompression is prevented. Because residual stresses do not depend on any external influences or on time, this kind of design is extremely simple.

5. **Recommendations for the design of non-exposed glass surfaces.** If a glass element's surfaces are permanently and safely protected from damage, and failure away from the edges is relevant, an RSFP-based model is suitable. *GlassTools*, the computer software that was developed as part of the present research, simplifies lifetime prediction and structural design with this model considerably. Nevertheless, the task may become difficult and time-consuming under complex loading conditions. In these cases, design flaw-based modelling is a useful, conservative and much simpler alternative.

6. **Recommendations in the case of machining damage.** Since the orientation and, more importantly, the location of the flaws are often not random, an RSFP-based model could produce unsafe results if glass elements contain significant machining damage. Such elements should, therefore, be designed using a design flaw that accounts for both machining damage and surface damage hazard scenarios.

7. **Recommendations for testing.** Full-scale testing remains an integral part of the design process of innovative glass structures. The present study has shown that the following should be observed:

- Results from laboratory testing at ambient conditions represent a combination of both surface condition and time-dependent crack growth. If testing at ambient conditions is unavoidable, subcritical crack growth during the tests must be modelled. While this can efficiently be done using the model and the software tools presented in this thesis, the dependence of the crack velocity parameters on the environmental conditions and the stress rate still diminishes the accuracy and reliability of the results. The problem can be addressed by the near-inert testing procedure that was developed and used in this thesis.
- Surface flaws have a decisive influence on the fracture strength. The structural performance of structural elements that may be exposed to severe surface damage should, therefore, be assessed by experiments on specimens with realistic design flaws rather than on as-received or homogeneously damaged specimens. Suitable tools and testing procedures were developed in this thesis.

8. **Limitations.** Complex models, such as the lifetime prediction model described in this thesis, are very helpful in gaining an in-depth understanding of the structural behaviour of glass elements and they are likely to foster further progress in research. The construction industry can benefit from such models for the interpretation of laboratory tests and for structural design. It is, however, important to be aware of the limitations with regard to structural design:

   - It was shown in this thesis that the crack velocity parameters vary very widely and depend on various external influences. Prediction of the crack growth over a service life of many years, and thereby of the design fracture strength, will inevitably be of limited accuracy.
   - However good a model for structural glass elements is, the accuracy of its output depends on a close match between the design surface damage and the actual in-service damage. The latter can only be estimated roughly.

   Consulting engineers, as well as researchers who provide them with guidance, must be aware of these limitations. It is preferable to perform structural glass design using simple methods that account for the key phenomena and are truly understood by their users rather than to use a complex model that has a high risk of being applied incorrectly and may offer unrealistic precision and a false sense of safety. Structural design based on the prevention of surface decompression, for instance, is such a simple method.

The key benefits of the present work can be summarized as follows:

- **Analysis of present knowledge.** A comprehensive analysis of present knowledge of the structural design of glass elements is provided. For the first time, the bases, underlying hypotheses, advantages, limits and shortcomings of current glass design methods are presented and discussed in detail.
- **Lifetime prediction model.** The lifetime prediction model can be used for predictive modelling and structural design of glass elements. It can furthermore be very helpful in gaining an in-depth understanding of the structural behaviour of such elements and has the potential to foster further scientific progress in the field. While the lifetime prediction model is more complex than traditional semi-empirical models, it offers significant advantages.
- **Mechanical behaviour of deep surface flaws.** The analytical and experimental investigations presented shed light on the mechanical behaviour of deep surface flaws in soda lime silica glass.
- **Improved testing procedures.** The laboratory testing procedures that are proposed and tested in this thesis increase the accuracy and reliability of experimental data and enable the engineer to obtain relevant data for structural design based on surface damage hazard scenarios.
- **Recommendations for practice.** Recommendations and useful information for engineering practice are provided.
- **Computer software.** The computer software 'GlassTools' allows calculations and simulations based on the models and concepts discussed in this thesis to be performed conveniently and quickly.

9.4 Further research

Based on the findings of this thesis, a few directions for future research are suggested:

- **Residual stress.** Quality control measures that allow for the use of a high design value for the residual surface compression stress of heat strengthened or fully tempered glass were found to be very efficient in terms of economical material use. In that respect, progress in at least two areas would be very beneficial:
    - Currently, glass manufacturers perform only extremely basic residual stress monitoring, for instance by shattering a glass pane per day to obtain an approximation of the residual stress level from the fragment size. The main reason for this is that precise surface stress measurements are very time-consuming. In future, engineers should be able to specify a minimum residual surface compression stress that the manufacturer has to guarantee. This requires fast and reliable surface compression measurement equipment. Furthermore, product codes need to adopt the concept of 'glass grading'.
    - The residual stress near edges and holes is not only different from the residual stress away from edges but also much more difficult to measure. It can, however, be determined quite accurately by numerical modelling of the tempering process. Past studies provided much insight, but the uncertainty and, consequently, the partial safety factor that must be applied to the residual stress for design, are still high. At least for the more common glass geometries, systematic studies that are backed by experiments should provide more reliable and comprehensive information. This would enable a lower partial safety factor and thereby a higher design value of the residual surface compression stress to be used.

- **Surface protection.** Since surface damage causes a substantial drop in the strength of glass elements, an active protection of the surface would enable safer and, by virtue of reduced element thickness and size, more elegant structures to be built. Such protection might, for instance, be achieved by applying a coating or plastic film to the glass surface. Chemical or mechanical polishing of the surface prior to coating could further reduce flaws and increase strength.

- **Surface damage hazard scenarios and design flaw properties.** When designing structural glass elements using the design flaw approach, the engineer needs to make a sensible assumption of the design damage that is caused by various surface damage hazards. The present work only gives an approximation of the order of magnitude of such damage. To limit the need for costly project specific testing as much as possible, experimental investigations should be conducted in order to establish a relationship between common hazard scenarios and the surface damage that they cause. These investigations should focus not only on the depth of the flaws, but also on their geometry factor.

- **Machining damage.** In order to predict the edge strength of glass elements and the strength at bolt holes accurately, the surface condition at these locations needs further investigation. It depends essentially on the machining process. It would, therefore, be very useful to determine the appropriate design flaw depth for common machining processes.

- **Design aids.** Although calculations using the lifetime prediction model are simplified and sped up considerably by the computer software that was developed as part of this research, they remain relatively complex. It would therefore be useful to develop design aids for practical application, such as tables and graphs for the most common design cases. Another more convenient and flexible approach would be to integrate *GlassTools'* algorithms into a widely used finite element package.

References

ABAQUS 2004. ABAQUS v6.5 (finite element software). Abaqus, Inc., USA, 2004. URL http://www.abaqus.com/.

Abrams, M. B., Green, D. J. and Glass, S. J. Fracture behavior of engineered stress profile soda lime silicate glass. *Journal of Non-Crystalline Solids*, 321(1-2):10–19, 2003.

ASTM C 1048-04. *Standard Specification for Heat-Treated Flat Glass – Kind HS, Kind FT Coated and Uncoated Glass*. American Society for Testing Materials, 2004.

ASTM E 1300-03. *Standard Practice for Determining Load Resistance of Glass in Buildings*. American Society for Testing Materials, 2003.

ASTM E 1300-04. *Standard Practice for Determining Load Resistance of Glass in Buildings*. American Society for Testing Materials, 2004.

ASTM E 1300-94. *Standard practice for determining the minimum thickness and type of glass required to resist a specific load*. American Society for Testing Materials, 1994.

Atkins, A. G. and Mai, Y.-W. *Elastic and Plastic Fracture*. Ellis Horwood, 1988. ISBN 0-132-48196-0.

Auradou, H., Vandembroucq, D., Guillot, C. and Bouchaud, E. A probabilistic model for the stress corrosion fracture of glass. *Transactions, SMiRT 16, Washington DC*, August 2001.

Bakioglu, M., Erdogan, F. and Hasselman, D. P. H. Fracture mechanical analysis of self-fatigue in surface compression strengthened glass plates. *Journal of Materials Science*, 11:1826–1834, 1976.

Bando, Y., Ito, S. and Tomozawa, M. Direct Observation of Crack Tip Geometry of SiO2 Glass by High-Resolution Electron Microscopy. *Journal of the American Ceramic Society*, 67(3):C36–C37, 1984.

Batdorf, S. B. and Crose, J. G. Statistical Theory for the Fracture of Brittle Structures Subjected to Nonuniform Polyaxial Stresses. *Journal of Applied Mechanics, Transactions ASME*, 41 Ser E(2):459–464, 1974.

Batdorf, S. B. and Heinisch, H. L. Weakest Link Theory Reformulated for Arbitrary Fracture Criterion. *Journal of the American Ceramic Society*, 61(7-8):355–358, 1978.

Beason, W. L. *A failure prediction model for window glass*. NTIS Accession no. PB81-148421, Texas Tech University, Institute for Disaster Research, 1980.

Beason, W. L., Kohutek, T. L. and Bracci, J. M. Basis for ASTM E 1300 annealed glass thickness selection charts. *Journal of Structural Engineering*, 142(2):215–221, February 1998.

Beason, W. L. and Morgan, J. R. Glass Failure Prediction Model. *Journal of Structural Engineering*, 110(2):197–212, 1984.

Bennison, S. J., Jagota, A. and Smith, C. A. Fracture of Glass/Poly(vinyl butyral) (Butacite®) Laminates in Biaxial Flexure. *Journal of the American Ceramic Society*, 82(7):1761–1770, 1999.

Bennison, S. J., Smith, C. A., Van Duser, A. and Jagota, A. Structural Performance of Laminated Glass Made with a "Stiff" Interlayer. In *Glass in Buildings, ASTM STP 1434, V. Block (editor)*. 2002.

Bernard, F. *Sur le dimensionnement des structures en verre trempé: étude des zones de connexion*. Ph.D. thesis, LMT Cachan, France, 2001.

REFERENCES

Bernard, F., Gy, R. and Daudeville, L. Finite Element Computation of Residual Stresses Near Holes in Tempered Glass Plates. *Glass Technology*, 43C, 2002.

Blank, K. *Dickenbemessung von vierseitig gelagerten rechteckigen Glasscheiben unter gleichförmiger Flächenlast*. Technical Report, Institut für Konstruktiven Glasbau, Gelsenkirchen, 1993.

Brückner-Foit, A., Munz, D., Schirmer, K. and Fett, T. Discrimination of multiaxiality criteria with the Brazilian disc test. *Journal of the European Ceramic Society*, 17(5):689–696, 1997.

BRL-A 2005. *Bauregelliste A.* German building regulation, Deutsches Institut für Bautechnik (DIBt), 2005.

Brown, W. G. A Load Duration Theory for Glass Design. *Publication NRCC 12354*, 1972.

Brown, W. G. *A practicable Formulation for the Strength of Glass and its Special Application to Large Plates*. Publication No. NRC 14372, National Research Council of Canada, Ottawa, 1974.

Bucak, O. and Ludwig, J. J. Stand der Untersuchungen an grossformatigen TVG-Scheiben zur Erlangung einer bauaufsichtlichen Zulassung. In *Bauen mit Glas - VDI-Tagung in Baden-Baden 2000*, pp. 677–706. VDI-Verlag, Düsseldorf, 2000.

Bunker, B. C. Molecular mechanisms for corrosion of silica and silicate glasses. *Journal of Non-Crystalline Solids*, 179:300–308, November 1994.

Calderone, I. The Dangers of Using a Probabilistic Approach for Glass Design. In *Proceedings of Glass Processing Days 2005, 17-20 June, Tampere, Finland*, pp. 364–366. 2005.

Calderone, I. J. The Fallacy of the Weibull Distribution for Window Glass Design. In *Proceedings of Glass Processing Days 2001, 18-21 June, Tampere, Finland*, pp. 293–297. 2001.

CAN/CGSB 12.20-M89. *Structural Design of Glass for Buildings*. Canadian General Standards Board, December 1989.

Charles, R. J. and Hilling, W. B. The kinetics of glass failure by stress corrosion. In *Symposium on Mechanical Strength of Glass and Ways of Improving it, Charleroy, Belgium*. 1962.

Charles, Y., Hild, F. and Roux, S. Long-term reliability of brittle materials: The issue of crack arrest. *Journal of Engineering Materials and Technology, Transactions of the ASME*, 125(3):333–340, 2003.

Choi, S. R. and Holland, F. Dynamic Fatigue Behavior of Soda-lime Glass Disk Specimens under Biaxial Loading, 1994. Unpublished work.

Choi, S. R., Salem, J. A. and Holland, F. A. *Estimation of Slow Crack Growth Parameters for Constant Stress-Rate Test Data of Advanced Ceramics and Glass by the Individual Data and Arithmetic Mean Methods*. NASA Technical Memorandum 107369, 1997.

Dalgliesh, W. A. and Taylor, D. A. The strength and testing of window glass. *Canadian Journal of Civil Engineering, Canada*, 17:752–762, 1990.

DIN 1249-10:1990. *Flachglas im Bauwesen – Teil 10: Chemische und physikalische Eigenschaften*. DIN, 1990.

DIN 18516-4:1990. *Außenwandbekleidungen, hinterlüftet – Einscheiben-Sicherheitsglas – Teil 4: Anforderungen, Bemessung, Prüfung*. DIN, 1990.

DIN 52292-2:1986. *Teil 2: Prüfung von Glas und Glaskeramik. Bestimmung der Biegefestigkeit, Doppelring-Biegeversuch an plattenförmigen Proben mit grossen Prüfflächen*. DIN, 1986.

DIN 55303-7:1996. *Statistische Auswertung von Daten – Teil 7: Schätz- und Testverfahren bei zweiparametriger Weibull-Verteilung*. DIN, 1996.

DuPont 2003. *SentryGlas Plus – The Interlayer for Structural Laminated Glass*. DuPont, 2003.

Dwivedi, P. and Green, D. J. Determination of Subcritical Crack Growth Parameters by In Situ Observations of Identation Cracks. *Journal of the American Ceramic Society*, 78(8):2122–2128, 1995.

EN 1051-1:2003. *Glass in building – Glass blocks and glass paver units – Part 1: Definitions and description*. CEN, 2003.

EN 1051-2:2003. *Glass in building – Glass blocks and glass paver units – Part 2: Evaluation of conformity*. CEN, 2003.

EN 1096-1:1998. *Glass in building – Coated glass – Part 1: Definitions and classification*. CEN, January 1999.

EN 1096-2:2001. *Glass in building – Coated glass – Part 2: Requirements and test methods for class A, B and S coatings*. CEN, 2001.

EN 1096-3:2001. *Glass in building – Coated glass – Part 3: Requirements and test methods for class C and D coatings*. CEN, 2001.

EN 1096-4:2004. *Glass in building – Coated glass – Part 4: Evaluation of conformity / Product standard*. CEN, 2004.

EN 12150-1:2000. *Glass in building – Thermally toughened soda lime silicate safety glass – Part 1: Definition and description.* CEN, 2000.

EN 12150-2:2004. *Glass in building – Thermally toughened soda lime silicate safety glass – Part 2: Evaluation of conformity / Product standard.* CEN, 2004.

EN 12337-1:2000. *Glass in building – Chemically strengthened soda lime silicate glass – Part 1: Definition and description.* CEN, 2000.

EN 12337-2:2004. *Glass in building – Chemically strengthened soda lime silicate glass – Part 2: Evaluation of conformity / Product standard.* CEN, 2004.

EN 12603:2002. *Glass in building – Procedures for goodness of fit and confidence intervals for Weibull distributed glass stength data.* CEN, 2002.

EN 1279-1:2004. *Glass in building – Insulating glass units – Part 1: Generalities, dimensional tolerances and rules for the system description.* CEN, 2004.

EN 1279-2:2002. *Glass in building – Insulating glass units – Part 2: Long term test method and requirements for moisture penetration.* CEN, 2002.

EN 1279-3:2002. *Glass in building – Insulating glass units – Part 3: Long term test method and requirements for gas leakage rate and for gas concentration tolerances.* CEN, 2002.

EN 1279-4:2002. *Glass in building – Insulating glass units – Part 4: Methods of test for the physical attributes of edge seals.* CEN, 2002.

EN 1279-5:2005. *Glass in building – Insulating glass units – Part 5: Evaluation of conformity.* CEN, 2005.

EN 1279-6:2002. *Glass in building – Insulating glass units – Part 6: Factory production control and periodic tests.* CEN, 2002.

EN 1288-1:2000. *Glass in building – Determination of the bending strength of glass – Part 1: Fundamentals of testing glass.* CEN, 2000.

EN 1288-2:2000. *Glass in building – Determination of the bending strength of glass – Part 2: Coaxial double ring test on flat specimens with large test surface areas.* CEN, 2000.

EN 1288-3:2000. *Glass in building – Determination of the bending strength of glass – Part 3: Test with specimen supported at two points (four point bending).* CEN, 2000.

EN 1288-5:2000. *Glass in building – Determination of the bending strength of glass – Part 5: Coaxial double ring test on flat specimens with small test surface areas.* CEN, 2000.

EN 13024-1:2002. *Glass in building – Thermally toughened borosilicate safety glass – Part 1: Definition and description.* CEN, 2002.

EN 13024-2:2004. *Glass in building – Thermally toughened borosilicate safety glass – Part 2: Evaluation of conformity / Product standard.* CEN, 2004.

EN 14178-1:2004. *Glass in building – Basic alkaline earth silicate glass products – Part 1: Float glass.* CEN, 2004.

EN 14178-2:2004. *Glass in building – Basic alkaline earth silicate glass products – Part 2: Evaluation of conformity / Product standard.* CEN, 2004.

EN 14179-1:2005. *Glass in building – Heat soaked thermally toughened soda lime silicate safety glass – Part 1: Definition and description.* CEN, 2005.

EN 14179-2:2005. *Glass in building – Heat soaked thermally toughened soda lime silicate safety glass – Part 2: Evaluation of conformity / Product standard.* CEN, 2005.

EN 14321-1:2005. *Glass in building – Thermally toughened alkaline earth silicate safety glass – Part 1: Definition and description.* CEN, 2005.

EN 14321-2:2005. *Glass in building – Thermally toughened alkaline earth silicate safety glass – Part 2: Evaluation of conformity / Product standard.* CEN, 2005.

EN 14449:2005. *Glass in building – Laminated glass and laminated safety glass – Evaluation of conformity / Product standard.* CEN, 2005.

EN 1748-1-1:2004. *Glass in building – Special basic products – Borosilicate glasses – Part 1-1: Definitions and general physical and mechanical properties.* CEN, 2004.

EN 1748-1-2:2004. *Glass in building – Special basic products – Borosilicate glasses – Part 1-2: Evaluation of conformity / Product standard.* CEN, 2004.

EN 1748-2-1:2004. *Glass in building – Special basic products – Glass ceramics – Part 2-1 Definitions and general physical and mechanical properties.* CEN, 2004.

EN 1748-2-2:2004. *Glass in building – Special basic products – Part 2-2: Glass ceramics – Evaluation of conformity / Product standard.* CEN, 2004.

EN 1863-1:2000. *Glass in building – Heat strengthened soda lime silicate glass – Part 1: Definition and description.* CEN, 2000.

EN 1863-2:2004. *Glass in building – Heat strengthened soda lime silicate glass – Part 2: Evaluation of conformity / Product standard.* CEN, 2004.

EN 1990:2002. *Grundlagen der Tragwerksplanung.* CEN, 2002.

EN 1991-1-1:2002. *Eurocode 1: Einwirkungen auf Tragwerke; Teil 1-1: Allgemeine Einwirkungen auf Tragwerke.* CEN, 2002.

EN 1991-2-3:1996. *Eurocode 1: Einwirkungen auf Tragwerke; Teil 2-3: Schneelasten.* CEN, 1996.

EN 1991-2-4:1995. *Eurocode 1: Einwirkungen auf Tragwerke; Teil 2-4: Windlasten.* CEN, December 1996.

EN 1991-2-5:1997. *Eurocode 1: Einwirkungen auf Tragwerke; Teil 2-5: Temperatureinwirkungen.* CEN, January 1999.

EN 1991-2-7:1998. *Eurocode 1: Einwirkungen auf Tragwerke; Teil 2-7: Aussergewöhnliche Einwirkungen.* CEN, July 2000.

EN 572-1:2004. *Glass in building – Basic soda lime silicate glass products – Part 1: Definitions and general physical and mechanical properties.* CEN, 2004.

EN 572-2:2004. *Glass in building – Basic soda lime silicate glass products – Part 2: Float glass.* CEN, 2004.

EN 572-3:2004. *Glass in building – Basic soda lime silicate glass products – Part 3: Polished wire glass.* CEN, 2004.

EN 572-4:2004. *Glass in building – Basic soda lime silicate glass products – Part 4: Drawn sheet glass.* CEN, 2004.

EN 572-5:2004. *Glass in building – Basic soda lime silicate glass products – Part 5: Patterned glass.* CEN, 2004.

EN 572-6:2004. *Glass in building – Basic soda lime silicate glass products – Part 6: Wired patterned glass.* CEN, 2004.

EN 572-7:2004. *Glass in building – Basic soda lime silicate glass products – Part 7: Wired or unwired channel shaped glass.* CEN, 2004.

EN 572-8:2004. *Glass in building – Basic soda lime silicate glass products – Part 8: Supplied and final cut sizes.* CEN, 2004.

EN 572-9:2004. *Glass in building – Basic soda lime silicate glass products – Part 9: Evaluation of conformity / Product standard.* CEN, 2004.

EN 673:1997. *Glass in building – Determination of thermal transmittance (U value) – Calculation method.* CEN, 1997.

Evans, A. G. A General Approach for the Statistical Analysis of Multiaxial Fracture. *Journal of the American Ceramic Society*, 61(7-8):302–308, 1978.

Evans, A. G. and Wiederhorn, S. M. Proof testing of ceramic materials – an analytical basis for failure prediction. *International Journal of Fracture*, 10(3):379–392, 1974.

Evans, A. G. and Wiederhorn, S. M. Proof testing of ceramic materials – an analytical basis for failure prediction. *International Journal of Fracture*, 26(4):355–368, 1984.

Evans, J. W., Johnson, R. A. and Green, D. W. *Two- and three-parameter Weibull goodness-of-fit tests.* Research Paper FPL-RP-493, U.S. Department of Agriculture, Forest Service, Forest Products Laboratory, 1989.

Exner, G. Erlaubte Biegespannung in Glasbauteilen im Dauerlastfall: Ein Vorhersagekonzept aus dynamischen Labor-Festigkeitsmessungen. *Glastechnische Berichte*, 56(11):299–312, 1983.

Exner, G. Abschätzung der erlaubten Biegespannung in vorgespannten Glasbauteilen. *Glastechnische Berichte*, 59(9):259–271, 1986.

Faber, M. *Risk and Safety in Civil Engineering.* Lecture Notes, Swiss Federal Institute of Technology ETH, Zürich, 2003.

Fink, A. *Ein Beitrag zum Einsatz von Floatglas als dauerhaft tragender Konstruktionswerkstoff im Bauwesen.* Ph.D. thesis, Technische Hochschule Darmstadt, 2000.

Fischer-Cripps, A. C. and Collins, R. E. Architectural glazings: Design standards and failure models. *Building and Environment*, 30(1):29–40, 1995.

GaertnerCorp 2002. *Stress Measuring Device Technical Manual.* Gaertner Scientific Corporation (Illinois, USA), 2002.

Gehrke, E., Ullner, C. and Hähnert, M. Effect of corrosive media on crack growth of model glasses and commercial silicate glasses. *Glastechnische Berichte*, 63(9):255–265, 1990.

Gehrke, E., Ullner, C. and Hähnert, M. Fatigue limit and crack arrest in alkali-containing silicate glasses. *Journal of Materials Science*, 26:5445–5455, 1991.

Green, D. J., Dwivedi, P. J. and Sglavo, V. M. Characterization of subcritical crack growth in ceramics using indentation cracks. *British Ceramic Transactions*, 98(6):291–295, 1999.

Grenet, L. Mechanical Strength of Glass. *Enc. Industr. Nat. Paris*, 5(4):838–848, 1899.

Griffith, A. A. The Phenomena of Rupture and Flow in Solids. *Philosophical Transactions, Series A*, 221:163–198, 1920.

Güsgen, J. *Bemessung tragender Bauteile aus Glas*. Ph.D. thesis, RWTH Aachen / Shaker Verlag, 1998.

Guin, J.-P. and Wiederhorn, S. Crack growth threshold in soda lime silicate glass: role of holdtime. *Journal of Non-Crystalline Solids*, 316(1):12–20, 2003.

Gy, R. Stress corrosion of silicate glass: a review. *Journal of Non-Crystalline Solids*, 316(1):1–11, 2003.

Haldimann, M., Luible, A. and Overend, M. *Structural use of Glass*. Structural Engineering Document SED. International Association for Bridge and Structural Engineering IABSE, Zürich (to be published), 2007.

Hand, R. J. Stress intensity factors for surface flaws in toughened glasses. *Fatigue and Fracture of Engineering Materials and Structures*, 23(1):73–80, 2000.

Hess, R. *Glasträger*. Forschungsbericht des Instituts für Hochbautechnik, vdf Hochschulverlag AG der ETH Zürich, Zürich, 2000.

Hénaux, S. and Creuzet, F. Kinetic fracture of glass at the nanometer scale. *Journal of Materials Science Letters*, 16(12):1008–1011, 1997.

Irwin, G. Analysis of Stresses and Strains near the End of a Crack Traversing a Plate. *Journal of Applied Mechanics*, 24:361–364, 1957.

Irwin, G. R. Crack Extension Force for a Part-Through Crack in a Plate. *Journal of Applied Mechanics*, 29(4):651–654, 1962.

Irwin, G. R., Krafft, J. M., Paris, P. C. and Wells, A. A. *Basic Aspects of Crack Growth and Fracture*. NRL Report 6598, Naval Research Laboratory, Washington D.C., 1967.

ISO 12543-1:1998. *Glass in building – Laminated glass and laminated safety glass – Part 1: Definitions and description of component parts*. ISO, 1998.

ISO 12543-2:2004. *Glass in building – Laminated glass and laminated safety glass – Part 2: Laminated safety glass*. ISO, 2004.

ISO 12543-3:1998. *Glass in building – Laminated glass and laminated safety glass – Part 3: Laminated glass*. ISO, 1998.

ISO 12543-4:1998. *Glass in building – Laminated glass and laminated safety glass – Part 4: Test methods for durability*. ISO, 1998.

ISO 12543-5:1998. *Glass in building – Laminated glass and laminated safety glass – Part 5: Dimensions and edge finishing*. ISO, 1998.

ISO 12543-6:1998. *Glass in building – Laminated glass and laminated safety glass – Part 6: Appearance*. ISO, 1998.

ISO 31-0:1992. *Quantities and units – Part 0: General principles*. ISO, 1992.

ISO 31-0:1992/Amd.2:2005. *Quantities and units – Part 0: General principles – Amendment 2*. ISO, 2005.

ISO 31-11:1992. *Quantities and units – Part 11: Mathematical signs and symbols for use in the physical sciences and technology*. ISO, 1992.

ISO 31-3:1992. *Quantities and units – Part 3: Mechanics*. ISO, 1992.

ISO/IEC 23270:2003. *Information technology – C# Language Specification*. ISO, 2003.

ISO/IEC 23271:2003. *Information technology – Common Language Infrastructure*. ISO, 2003.

Jacob, L. and Calderone, I. J. A New Design Model Based on Actual Behaviour of Glass Panels Subjected to Wind. In *Proceedings of Glass Processing Days 2001, 18-21 June, Tampere, Finland*, pp. 311–314. 2001.

Johar, S. *Dynamic fatigue of flat glass – Phase II*. Technical Report, Ontario Research Foundation. Department of Metals, Glass and Ceramics, Missisagua, Canada, 1981.

Johar, S. *Dynamic fatigue of flat glass – Phase III*. Technical Report, Ontario Research Foundation. Department of Metals, Glass and Ceramics, Missisagua, Canada, 1982.

Kerkhof, F. Bruchmechanische Analyse von Schadensfällen an Gläsern. *Glastechnische Berichte*, 1975.

REFERENCES

Kerkhof, F. Bruchmechanik von Glas und Keramik. *Sprechsaal*, 110:392–397, 1977.

Kerkhof, F., Richter, H. and Stahn, D. Festigkeit von Glas – Zur Abhängigkeit von Belastungsdauer und -verlauf. *Glastechnische Berichte*, 54(8):265–277, 1981.

Kott, A. *Zum Trag- und Resttragverhalten von Verbundsicherheitsglas*. Ph.D. thesis, Eidgenössische Technische Hochschule Zürich (ETH) - Institut für Baustatik und Konstruktion (IBK), Zurich, Switzerland, 2006.

Kurkjian, C. R., Gupta, P. K., Brow, R. K. and Lower, N. The intrinsic strength and fatigue of oxide glasses. *Journal of Non-Crystalline Solids*, 316(1):114–124, 2003.

Laufs, W. *4. Forschungsbericht Fachverband Konstruktiver Glasbau, Arbeitskreis Punktgestützte Gläser*. Technical Report, Fachverband Konstruktiver Glasbau, Arbeitskreis Punktgestützte Gläser, Aachen, 2000a.

Laufs, W. *Ein Bemessungskonzept zur Festigkeit thermisch vorgespannter Gläser*. Ph.D. thesis, RWTH Aachen / Shaker Verlag, 2000b.

Lawn, B. *Fracture of brittle solids*. Second Edition. Cambridge University Press, 1993.

Levengood, W. C. Effect of origin flaw characteristics on glass strength. *Journal of Applied Physics*, 29(5):820–826, 1958.

Lotz, S. *Untersuchung zur Festigkeit und Langzeitbeständigkeit adhäsiver Verbindungen zwischen Fügepartnern aus Floatglas*. Ph.D. thesis, Universität Kaiserslautern, Fachbereich Maschinenwesen, 1995.

Luible, A. *Knickversuche an Glasscheiben*. Rapport ICOM 479, Ecole polytechnique fédérale de Lausanne (EPFL), 2004a.

Luible, A. *Stabilität von Tragelementen aus Glas*. Thèse EPFL 3014, Ecole polytechnique fédérale de Lausanne (EPFL), 2004b. URL http://icom.epfl.ch/publications/pubinfo.php?pubid=499.

Marlière, C., Prades, S., Célarié, F., Dalmas, D., Bonamy, D., Guillot, C. and Bouchaud, E. Crack fronts and damage in glass at the nanometre scale. *Journal of Physics: Condensed Matter*, 15:2377–2386, 2003.

Matlab 2005. Matlab 7.1 (technical computing software). The MathWorks, Inc., 2005. URL http://www.mathworks.com/products/matlab/.

Matlab Doc 2005. *Documentation for Matlab 7.1*. The MathWorks, Inc., 2005.

Menčík, J. Rationalized Load and Lifetime of Brittle Materials. *Journal of the American Ceramic Society*, 67(3):C37–C40, 1984.

Menčík, J. *Strength and Fracture of Glass and Ceramics*, Glass Science and Technology, volume 12. Elsevier, 1992.

Michalske, T. A. and Freiman, S. W. A Molecular Mechanism for Stress Corrosion in Vitreous Silica. *Journal of the American Ceramic Society*, 66(4):284–288, 1983.

MS Excel 2003. Microsoft Excel 2003 (software). Microsoft Corporation, One Microsoft way, Redmond, USA, 2003.

Nelson, W. B. *Applied Life Data Analysis*. John Wiley & Sons, Inc., 2003. ISBN 0-471-64462-5.

.NET Framework website. Microsoft .NET Framework Developer Center. 2006. URL http://msdn.microsoft.com/netframework/.

Newman, J. C. and Raju, I. S. Analysis of Surface Cracks in Finite Plates Under Tension or Bending Loads. *NASA Technical Paper 1578*, 1979.

Nghiem, B. *Fracture du verre et heterogeneites a l'echelle sub-micronique*. Ph.D Thesis Nr 98 PA06 6260, Université Paris-VI, 1998.

Overend, M. *The Appraisal of Structural Glass Assemblies*. Ph.D. thesis, University of Surrey, March 2002.

Pilkington Planar 2005. *Pilkington Planar*. Product documentation, W&W Glass, USA, 2005. URL http://www.wwglass.com/pdf/2005Catalog.pdf.

Porter, M. *Aspects of Structural Design with Glass*. Ph.D. thesis, University of Oxford, 2001.

prEN 13474-1:1999. *Glass in building – Design of glass panes – Part 1: General basis of design*. CEN, 1999.

prEN 13474-2:2000. *Glass in building – Design of glass panes – Part 2: Design for uniformly distributed load*. CEN, 2000.

Reid, S. G. Flaws in the failure prediction model of glass strength. In *Proceedings of the 6th International Conference on Applications of Statistics and Probability in Civil Engineering, Mexico City*, pp. 111–117. 1991.

Reid, S. G. Model Errors in the Failure Prediction Model of Glass Strength. In *8th ASCE Specialty Conference on Probabilistic Mechanics and Structural Reliability, Alberta, Canada*. 2000.

Richter, H. *Experimentelle Untersuchungen zur Rissausbreitung in Spiegelglas im Geschwindigkeitsbereich 10E-3 bis 5 10E3 mm/s*. Ph.D. thesis, Institut für Festkörpermechanik Freiburg, 1974.

Ritter, J. E., Service, T. H. and Guillemet, C. Strength an fatigue parameters for soda-lime glass. *Glass Technology*, 26(6):273–278, December 1985.

Ritter, J. E., Strzepa, P., Jakus, K., Rosenfeld, L. and Buckman, K. J. Erosion Damage in Glass and Alumina. *Journal of the American Ceramic Society*, 67(11):769–774, 1984.

Sayir, M. B. *Mechanik 2 – Deformierbare Körper*. ETH Zürich, 1996.

Schmitt, R. W. *Entwicklung eines Prüfverfahrens zur Ermittlung der Biegefestigkeit von Glas und Aspekte der statistischen Behandlung der gewonnenen Messwerte*. Ph.D. thesis, RWTH Aachen, 1987.

Schneider, J. *Festigkeit und Bemessung punktgelagerter Gläser und stossbeanspruchter Gläser*. Ph.D. thesis, TU Darmstadt, Institut für Statik, 2001.

Sedlacek, G., Blank, K., Laufs, W. and Güsgen, J. *Glas im Konstruktiven Ingenieurbau*. Ernst & Sohn, Berlin, 1999. ISBN 3-433-01745-X.

Semwogerere, D. and Weeks, E. R. *Confocal Microscopy*, chapter to be published in the Encyclopedia of Biomaterials and Biomedical Engineering. Taylor & Francis, 2005.

Sglavo, V. and Bertoldi, M. Vickers indentation: A powerful tool for the analysis of fatigue behavior on glass. *Ceramic Transactions*, 156:13–22, 2004.

Sglavo, V., Gadotti, M. and Micheletti, T. Cyclic loading behaviour of soda lime silicate glass using indentation cracks. *Fatigue and Fracture of Engineering Materials and Structures*, 20(8):1225–1234, 1997.

Sglavo, V., Prezzi, A. and Zandonella, T. Engineered stress-profile silicate glass: High strength material insensitive to surface defects and fatigue. *Advanced Engineering Materials*, 6(5):344–349, 2004.

Sglavo, V. M. ESP (Engineered Stress Profile) Glass: High Strength Material, Insensitive to Surface Defects. In *Proceedings of Glass Processing Days 2003*, 15-18 June, Tampere, Finland, pp. 74–77. 2003.

Sglavo, V. M. and Green, D. J. Subcritical growth of indentation median cracks in soda-lime-silica glass. *Journal of the American Ceramic Society*, 78(3):650–656, 1995.

Sglavo, V. M. and Green, D. J. Indentation fatigue testing of soda-lime silicate glass. *Journal of Materials Science*, 34(3):579–585, 1999.

Shen, X. *Entwicklung eines Bemessungs- und Sicherheitskonzeptes für den Glasbau*. Ph.D. thesis, Technische Hochschule Darmstadt, 1997.

SIA 260:2003 (e). *SIA 260: Basis of structural design*. Schweizerischer Ingenieur- und Architektenverein, Zürich, 2003.

SIA 261:2003 (e). *SIA 261: Actions on Structures*. Schweizerischer Ingenieur- und Architektenverein, Zürich, 2003.

Siebert, G. *Beitrag zum Einsatz von Glas als tragendes Bauteil im konstruktiven Ingenieurbau*. Ph.D. thesis, Technische Universität München (TUM), 1999.

Siebert, G. *Entwurf und Bemessung von tragenden Bauteilen aus Glas*. Ernst & Sohn, Berlin, 2001. ISBN 3-433-01614-3.

Simiu, E., Reed, A., Yancey, C. W. C., Martin, J. W., Hendrickson, E. M., Gonzalez, A. C., Koike, M., Lechner, J. A. and Batts, M. E. *Ring-on-ring tests and load capacity of cladding glass*. NBS Building Science Series 162, U.S. Department of Commerce - National Bureau of Standards, 1984.

Simmons, C. J. and Freiman, S. W. Effect of corrosion processes on subcritical crack growth in glass. *Journal of the American Ceramic Society*, 64, 1981.

START 2003:5. *Anderson-Darling: A goodness of fit test for small samples assumptions*. START 2003-5, Reliability Information Analysis Center RIAC, US Department of Defense, 2003.

Stavrinidis, B. and Holloway, D. G. Crack healing in Glass. *Physics and Chemistry of Glasses*, 24(1):19–25, 1983.

Stephens, M. A. EDF Statistics for Goodness of Fit and Some Comparisons. *Journal of the American Statistical Association*, 69(347):730–737, 1974.

Tada, H., Paris, P. C. and Irwin, G. R. *The Stress Analysis of Cracks Handbook*. Paris Productions Inc., 1985.

Thiemeier, T. *Lebensdauervorhersage für keramische Bauteile unter mehrachsiger Beanspruchung*. Ph.D. thesis, Universität Karlsruhe, 1989.

Thiemeier, T., Brückner-Foit, A. and Kölker, H. Influence of the fracture criterion on the failure prediction of ceramics loaded in biaxial flexure. *Journal of the American Ceramic Society*, 74(1):48–52, 1991.

Tomozawa, M. Stress corrosion reaction of silica glass and water. *Physics and Chemistry of Glasses*, 39(2):65–69, 1998.

TRAV 2003. *Technische Regeln für die Verwendung von absturzsichernden Verglasungen (TRAV)*. Technical Report, Deutsches Institut für Bautechnik (DIBt), January 2003.

TRLV 1998. *Technische Regeln für die Verwendung von linienförmig gelagerten Verglasungen (TRLV)*. Technical Report, Mitteilungen des Deutschen Instituts für Bautechnik (DIBt), Berlin, 1998.

Ullner, C. *Untersuchungen zum Festigkeitsverhalten von Kalk-Natronsilicatglas nach mechanischer Vorschädigung durch Korundberieselung*. Technical Report, BAM, Berlin, 1993.

Ullner, C. and Höhne, L. *Untersuchungen zum Festigkeitsverhalten und zur Rissalterung von Glas unter dem Einfluss korrosiver Umgebungsbedingungen*. Technical Report, BAM, Berlin, 1993.

Vallabhan, C. V. G. Iterative Analysis of Nonlinear Glass Plates. *Journal of Structural Engineering*, 109(2):489–502, 1983.

VS Express website. Microsoft Visual Studio Express website. 2006. URL http://msdn.microsoft.com/vstudio/express/.

Walker, G. R. and Muir, L. M. An Investigation of the Bending Strength of Glass Louvre Blades. In *Proceedings of the 9th Australian Conference on the Mechanics of Structures and Materials, Sydney, Australia*. 1984.

Weibull, W. A Statistical Theory of the Strength of Materials. *Ingeniors Vetenskaps Akademien (Proceedings of the Royal Swedish Academy of Engineering)*, 151, 1939.

Weibull, W. A Statistical Distribution Function of Wide Applicability. *Journal of Applied Mechanics*, 18:293–297, September 1951.

Wiederhorn, S., Dretzke, A. and Rödel, J. Crack growth in soda-lime-silicate glass near the static fatigue limit. *Journal of the American Ceramic Society*, 85(9):2287–2292, 2002.

Wiederhorn, S. and Tornsend, P. Crack healing in Glass. *Journal of the American Ceramic Society*, 53:486–489, 1970.

Wiederhorn, S. M. Influence of Water Vapor on Crack Propagation in Soda-Lime Glass. *Journal of the American Ceramic Society*, 50:407–414, 1967.

Wiederhorn, S. M. and Bolz, L. H. Stress corrosion and static fatigue of glass. *Journal of the American Ceramic Society*, 53(10):543–548, 1970.

Wiederhorn, S. M., Dretzke, A. and Rödel, J. Near the static fatigue limit in glass. *International Journal of Fracture*, 121(1-2):1–7, 2003.

Wiederhorn, S. M., Fuller Jr., E. R. and Thomson, R. Micromechanisms of crack growth in ceramics and glasses in corrosive environments. *Metal Science*, 14(8-9):450–458, 1980.

Wikipedia. Wikipedia – The Free Encyclopedia. 2006. URL http://www.wikipedia.org/.

Wörner, J.-D., Schneider, J. and Fink, A. *Glasbau: Grundlagen, Berechnung, Konstruktion*. Springer Verlag, 2001. ISBN 3-540-66881-0.

www.mono-project.com. Mono – An open-source project providing the necessary software to develop and run .NET client and server applications on Linux, Solaris, Mac OS X, Windows, and Unix. 2006. URL http://www.mono-project.com/.

www.weibull.com. On-line Reliability Engineering Resources. 2006. URL http://www.weibull.com/.

Appendices

Appendix

A

Overview of European Standards

Table A.1: Overview of European standards for basic glass products (shortened titles).

EN 572-1:2004	Basic soda lime silicate glass products – Part 1: Definitions and general physical and mechanical properties
EN 572-2:2004	Basic soda lime silicate glass products – Part 2: Float glass
EN 572-3:2004	Basic soda lime silicate glass products – Part 3: Polished wire glass
EN 572-4:2004	Basic soda lime silicate glass products – Part 4: Drawn sheet glass
EN 572-5:2004	Basic soda lime silicate glass products – Part 5: Patterned glass
EN 572-6:2004	Basic soda lime silicate glass products – Part 6: Wired patterned glass
EN 572-7:2004	Basic soda lime silicate glass products – Part 7: Wired or unwired channel shaped glass
EN 572-8:2004	Basic soda lime silicate glass products – Part 8: Supplied and final cut sizes
EN 572-9:2004	Basic soda lime silicate glass products – Part 9: Evaluation of conformity / Product standard
EN 1748-1-1:2004	Special basic products – Borosilicate glasses – Part 1-1: Definitions and general physical and mechanical properties
EN 1748-1-2:2004	Special basic products – Borosilicate glasses – Part 1-2: Evaluation of conformity / Product standard
EN 1748-2-1:2004	Special basic products – Glass ceramics – Part 2-1 Definitions and general physical and mechanical properties
EN 1748-2-2:2004	Special basic products – Glass ceramics – Part 2-2: Evaluation of conformity / Product standard.
EN 1051-1:2003	Glass blocks and glass paver units – Part 1: Definitions and description
EN 1051-2:2003	Glass blocks and glass paver units – Part 2: Evaluation of conformity
EN 14178-1:2004	Basic alkaline earth silicate glass products – Part 1: Float glass
EN 14178-2:2004	Basic alkaline earth silicate glass products – Part 2: Evaluation of conformity / Product standard

APPENDIX A. OVERVIEW OF EUROPEAN STANDARDS

Table A.2: Overview of European standards for processed glass products (shortened titles).

EN 1863-1:2000	Heat strengthened soda lime silicate glass – Part 1: Definition and description
EN 1863-2:2004	Heat strengthened soda lime silicate glass – Part 2: Evaluation of conformity / Product standard
EN 12150-1:2000	Thermally toughened soda lime silicate safety glass – Part 1: Definition and description
EN 12150-2:2004	Thermally toughened soda lime silicate safety glass – Part 2: Evaluation of conformity / Product standard
EN 14179-1:2005	Heat soaked thermally toughened soda lime silicate safety glass – Part 1: Definition and description
EN 14179-2:2005	Heat soaked thermally toughened soda lime silicate safety glass – Part 2: Evaluation of conformity / Product standard
EN 13024-1:2002	Thermally toughened borosilicate safety glass – Part 1: Definition and description
EN 13024-2:2004	Thermally toughened borosilicate safety glass – Part 2: Evaluation of conformity / Product standard
EN 14321-1:2005	Thermally toughened alkaline earth silicate safety glass – Part 1: Definition and description
EN 14321-2:2005	Thermally toughened alkaline earth silicate safety glass – Part 2: Evaluation of conformity / Product standard
EN 12337-1:2000	Chemically strengthened soda lime silicate glass – Part 1: Definition and description
EN 12337-2:2004	Chemically strengthened soda lime silicate glass – Part 2: Evaluation of conformity / Product standard
EN 1096-1:1998	Coated glass – Part 1: Definitions and classification
EN 1096-2:2001	Coated glass – Part 2: Requirements and test methods for class A, B and S coatings
EN 1096-3:2001	Coated glass – Part 3: Requirements and test methods for class C and D coatings
EN 1096-4:2004	Coated glass – Part 4: Evaluation of conformity / Product standard
ISO 12543-1:1998	Laminated glass and laminated safety glass – Part 1: Definitions and description of component parts
ISO 12543-2:2004	Laminated glass and laminated safety glass – Part 2: Laminated safety glass
ISO 12543-3:1998	Laminated glass and laminated safety glass – Part 3: Laminated glass
ISO 12543-4:1998	Laminated glass and laminated safety glass – Part 4: Test methods for durability
ISO 12543-5:1998	Laminated glass and laminated safety glass – Part 5: Dimensions and edge finishing
ISO 12543-6:1998	Laminated glass and laminated safety glass – Part 6: Appearance
EN 14449:2005	Laminated glass and laminated safety glass – Evaluation of conformity / Product standard
EN 1279-1:2004	Insulating glass units – Part 1: Generalities, dimensional tolerances and rules for the system description
EN 1279-2:2002	Insulating glass units – Part 2: Long term test method and requirements for moisture penetration
EN 1279-3:2002	Insulating glass units – Part 3: Long term test method and requirements for gas leakage rate and for gas concentration tolerances
EN 1279-4:2002	Insulating glass units – Part 4: Methods of test for the physical attributes of edge seals
EN 1279-5:2005	Insulating glass units – Part 5: Evaluation of conformity
EN 1279-6:2002	Insulating glass units – Part 6: Factory production control and periodic tests

Appendix B

On the Derivation of the Lifetime Prediction Model

Equation (4.76) gives the failure probability of a single flaw exposed to time-dependent loading and subcritical crack growth:

$$P_f^{(1)}(t) = \mathscr{P}\left(a_i \geq \min_{\tau \in [0,t]} \tilde{a}_c(\tau)\right) = \left(\frac{a_0}{\min_{\tau \in [0,t]} \tilde{a}_c(\tau)}\right)^{m_0/2} \tag{B.1}$$

Using the survival probability of an element with a mean number of flaws M from Equation (4.42) and $M = M_0/A_0 \cdot A$ from Equation (4.46), this equals

$$P_f(t) = 1 - \exp\left(-M \cdot P_f^{(1)}(t)\right) = 1 - \exp\left(-\frac{M_0}{A_0} A \cdot P_f^{(1)}(t)\right). \tag{B.2}$$

Multiplication of $P_f^{(1)}(t)$ by the probability $P_{\varphi,\vec{r}} = 1/A\,dA \cdot 1/\pi\,d\varphi$ of finding a crack of orientation φ at the point $\vec{r}$ on the surface (Equation (4.66)) and integration over all crack locations and orientations (cf. Section 4.3.3) yields

$$P_f(t) = 1 - \exp\left\{-\frac{1}{A_0}\int_A \frac{2}{\pi}\int_{\varphi=0}^{\pi/2} M_0 P_f^{(1)}(t,\vec{r},\varphi)\,dA\,d\varphi\right\}. \tag{B.3}$$

The term $M_0 P_f^{(1)}(t,\vec{r},\varphi)$ can now be obtained by inserting $\tilde{a}_c(\tau)$ from Equation (4.20) into Equation (B.1):

$$M_0 P_f^{(1)}(t,\vec{r},\varphi) = \left[\frac{a_0 \cdot M_0^{2/m_0}}{\min_{\tau \in [0,t]}\left\{\left(\left(\frac{\sigma_n(\tau,\vec{r},\varphi)Y\sqrt{\pi}}{K_{Ic}}\right)^{n-2} + \frac{n-2}{2}v_0 K_{Ic}^{-n}(Y\sqrt{\pi})^n \int_0^\tau \sigma_n^n(\tilde{\tau},\vec{r},\varphi)\,d\tilde{\tau}\right)^{\frac{2}{2-n}}\right\}}\right]^{\frac{m_0}{2}}$$

$$= \left[\sqrt{a_0}M_0^{1/m_0}\left\{\min_{\tau \in [0,t]}\left\{\left(\left(\frac{\sigma_n(\tau,\vec{r},\varphi)Y\sqrt{\pi}}{K_{Ic}}\right)^{n-2} + \frac{n-2}{2}v_0 K_{Ic}^{-n}(Y\sqrt{\pi})^n \int_0^\tau \sigma_n^n(\tilde{\tau},\vec{r},\varphi)\,d\tilde{\tau}\right)^{\frac{1}{2-n}}\right\}\right\}^{-1}\right]^{m_0}$$

$$= \left[\sqrt{a_0} M_0^{1/m_0} \cdot \max_{\tau \in [0,t]} \left\{ \left(\left(\frac{\sigma_n(\tau, \vec{r}, \varphi) Y \sqrt{\pi}}{K_{Ic}} \right)^{n-2} + \frac{n-2}{2} v_0 K_{Ic}^{-n} (Y \sqrt{\pi})^n \int_0^\tau \sigma_n^n(\tilde{\tau}, \vec{r}, \varphi) d\tilde{\tau} \right)^{\frac{1}{n-2}} \right\} \right]^{m_0} \quad (B.4)$$

Note that the minimum-function became a maximum-function because the exponent -1 was applied to its argument. The expression is now reformulated in order to be able to use the surface condition parameter

$$\theta_0 = \frac{K_{Ic}}{M_0^{1/m_0} \cdot Y \sqrt{\pi} \cdot \sqrt{a_0}}, \quad (B.5)$$

which was already used for the inert failure probability in Equation (4.48):

$$M_0 P_f^{(1)}(t, \vec{r}, \varphi) = \left[\max_{\tau \in [0,t]} \left\{ \left(\left(\sqrt{a_0} M_0^{1/m_0} \right)^{n-2} \left(\frac{\sigma_n(\tau, \vec{r}, \varphi)}{\frac{K_{Ic}}{Y\sqrt{\pi}}} \right)^{n-2} + \frac{n-2}{2} v_0 \left(\sqrt{a_0} M_0^{1/m_0} \right)^{n-2} \frac{1}{K_{Ic}^{n-2} K_{Ic}^2} (Y\sqrt{\pi})^{n-2} (Y\sqrt{\pi})^2 \int_0^\tau \sigma_n^n(\tilde{\tau}, \vec{r}, \varphi) d\tilde{\tau} \right)^{\frac{1}{n-2}} \right\} \right]^{m_0}$$

$$= \left[\max_{\tau \in [0,t]} \left\{ \left(\left(\frac{\sigma_n(\tau, \vec{r}, \varphi)}{\frac{K_{Ic}}{M_0^{1/m_0} Y \sqrt{\pi} \sqrt{a_0}}} \right)^{n-2} + \frac{(n-2) v_0 Y^2 \pi}{2 K_{Ic}^2} \left(\frac{M_0^{1/m_0} Y \sqrt{\pi} \sqrt{a_0}}{K_{Ic}} \right)^{n-2} \int_0^\tau \sigma_n^n(\tilde{\tau}, \vec{r}, \varphi) d\tilde{\tau} \right)^{\frac{1}{n-2}} \right\} \right]^{m_0}$$

$$= \left[\max_{\tau \in [0,t]} \left\{ \left(\left(\frac{\sigma_n(\tau, \vec{r}, \varphi)}{\theta_0} \right)^{n-2} + \frac{1}{U \cdot \theta_0^{n-2}} \cdot \int_0^\tau \sigma_n^n(\tilde{\tau}, \vec{r}, \varphi) d\tilde{\tau} \right)^{\frac{1}{n-2}} \right\} \right]^{m_0} \quad (B.6)$$

with the combined coefficient U containing the remaining parameters related to fracture mechanics and subcritical crack growth:

$$U = \frac{2 K_{Ic}^2}{(n-2) v_0 (Y \sqrt{\pi})^2} \quad (B.7)$$

Inserting this back into Equation (B.3) finally yields Equation (4.77).

Appendix

C

EPFL-ICOM Laboratory Testing Information and Data

C.1 Test data – As-received glass

Table C.1: Test series and parameter overview for tests on as-received annealed glass.

Test	Specimens	Average eff. load rate (kN/s)	Average eff. stress rate (MPa/s)	Test conditions	Ambient temperature (°C)	Relative humidity (%)
AS	10	0.013	0.21	ambient, slow	23.8	51.4
AF	10	1.32	21.2	ambient, fast	23.4	54.7
IS	10	0.013	0.21	dry/coated, slow	23.9	51.7
IF	10	1.35	21.6	dry/coated, fast	23.2	53.7
IV	4	2.46	39.4	dry/coated, very fast	23.4	54.7

Table C.2: Test data – As-received glass at ambient conditions (AS, AF).

Id	Thickness (mm)	Load rate (N/s)	Time to failure (s)	Force at failure (kN)	Stress at failure (MPa)
AS-01	5.85	13.0	198.68	2.580	42.2
AS-02	5.84	13.0	217.16	2.820	46.1
AS-03	5.84	12.8	258.41	3.304	53.9
AS-04	5.85	12.6	199.60	2.521	41.3
AS-05	5.87	13.2	282.20	3.729	60.8
AS-06	5.85	12.9	228.37	2.946	48.1
AS-07	5.87	13.2	338.81	4.458	72.5
AS-08	5.87	13.2	301.75	3.994	65.0
AS-09	5.87	13.0	161.04	2.086	34.2
AS-10	5.85	13.0	419.37	5.456	88.4
AF-01	5.83	1321	6.89	9.100	145.5
AF-02	5.83	1294	3.49	4.519	73.4
AF-03	5.83	1317	4.54	5.971	96.5
AF-04	5.83	1319	4.14	5.459	88.4
AF-05	5.85	1351	6.64	8.964	143.4
AF-06	5.86	1330	4.23	5.627	91.1
AF-07	5.87	1342	5.79	7.774	124.9
AF-08	5.84	1323	5.64	7.459	120.0
AF-09	5.84	1300	3.10	4.036	65.7
AF-10	5.85	1311	3.89	5.092	82.6

Table C.3: Test data – As-received glass at near-inert conditions (IS, IF, IV).

Id	Thickness (mm)	Load rate (N/s)	Time to failure (s)	Force at failure (kN)	Stress at failure (MPa)
IS-01	5.86	12.80	473.63	6.061	98.0
IS-02	5.86	12.79	269.68	3.450	56.3
IS-03	5.86	12.88	482.37	6.212	100.4
IS-04	5.88	12.92	604.60	7.812	125.5
IS-05	5.87	12.79	368.95	4.720	76.7
IS-06	5.85	12.90	352.66	4.549	73.9
IS-07	5.85	12.83	401.17	5.147	83.5
IS-08	5.85	12.55	392.50	4.927	80.0
IS-09	5.84	12.56	312.39	3.922	63.9
IS-10	5.85	12.92	446.11	5.762	93.2
IF-01	5.83	1338	6.59	8.814	141.1
IF-02	5.83	1349	7.66	10.338	164.6
IF-03	5.85	1338	6.34	8.486	136.0
IF-04	5.84	1324	4.71	6.232	100.7
IF-05	5.84	1370	8.79	12.045	190.6
IF-06	5.86	1353	7.73	10.466	166.5
IF-07	5.87	1370	8.01	10.966	174.2
IF-08	5.85	1345	6.64	8.929	142.9
IF-09	5.87	1365	7.67	10.473	166.7
IF-10	5.86	1335	5.91	7.894	126.8
IV-01	5.82	2479	4.57	11.319	179.6
IV-02	5.81	2439	2.93	7.154	115.2
IV-03	5.82	2487	4.94	12.291	194.3
IV-04	5.81	2423	3.19	7.742	124.4

C.2 Test data – Intentionally scratched glass

Table C.4: Test data - Intentionally scratched glass at near-inert conditions (SI.A, SI.H, SI.F).

Id	Thickness (mm)	σ_r (MPa)	Scratching device load (kg)	a_{s1} (μm)	a_{s2} (μm)	Effective load rate (N/s)	Effective stress rate (MPa/s)	t_f (s)	$Q(t_f)$ (kN)	$\sigma_E(t_f)$ (MPa)	$\sigma_n(t_f)$ (MPa)
SI.A-01	5.82		0.5	3	2	1248	19.7	4.89	6.100	98.6	98.6
SI.A-02	5.80		0.5	2	2	1323	21.5	1.93	2.558	41.9	41.9
SI.A-03	5.83		1.0	19	10	1182	19.2	1.91	2.258	37.0	37.0
SI.A-04	5.83		1.0	11	31	1197	19.3	2.72	3.252	53.1	53.1
SI.A-05	5.82		2.5	30	25	1328	21.5	2.10	2.784	45.5	45.5
SI.A-06	5.81		2.5	24	35	1016	16.5	2.27	2.306	37.8	37.8
SI.A-07	5.81		2.5	43	48	954	15.5	1.90	1.815	29.8	29.8
SI.A-08	5.80		2.5	28	54	1198	19.4	2.37	2.834	46.3	46.3
SI.A-09	5.81		2.5	10	9	956	15.6	1.75	1.670	27.4	27.4
SI.A-10	5.80		3.5	28	24	1173	19.1	1.89	2.222	36.4	36.4
SI.A-11	5.80		3.5	37	30	1054	17.1	2.33	2.458	40.2	40.2
SI.H-01	5.91	−71	0.5	3	7	1411	21.5	8.07	11.389	180.6	110.0
SI.H-02	5.88	−64	1.0	11	9	1249	19.4	6.95	8.674	138.9	74.7
SI.H-03	5.89	−67	1.0	3	33	1242	19.4	6.54	8.127	130.4	62.9
SI.H-04	5.89	−64	2.5	46	46	1392	22.0	4.20	5.850	94.6	30.4
SI.H-05	5.89	−71	2.5	29	36	1258	19.8	5.57	7.014	113.0	42.3
SI.H-06	5.91	−67	3.5	28	41	1276	20.0	5.93	7.568	121.7	54.2
SI.H-07	5.89	−71	3.5	43	16	1225	19.3	5.65	6.922	111.6	40.9
SI.F-01	5.91	−125	0.5	24	15	1420	21.5	8.47	12.023	190.3	65.0
SI.F-02	5.93	−125	0.5	7	10	1454	21.2	11.82	17.187	266.9	141.6
SI.F-03	5.92	−119	1.0	24	3	1299	19.1	12.10	15.711	245.3	126.4
SI.F-04	5.89	−122	1.0	10	3	1268	18.8	11.38	14.434	226.4	104.3
SI.F-05	5.92	−122	2.5	39	22	1431	21.6	8.46	12.110	191.6	69.5
SI.F-06	5.89	−122	2.5	28	27	1400	21.2	8.41	11.767	186.4	64.3
SI.F-07	5.91	−122	2.5	56	53	1256	19.4	7.30	9.171	146.6	24.5
SI.F-08	5.89	−122	2.5	43	8	1263	19.1	9.84	12.437	196.5	74.4
SI.F-09	5.91	−122	2.5	33	47	1276	19.2	10.06	12.836	202.5	80.4
SI.F-10	5.90	−129	3.5	15	34	1228	18.8	8.75	10.750	170.9	42.4
SI.F-11	5.88	−125	3.5	36	51	1245	19.1	8.31	10.346	164.7	39.4

Table C.5: Test data - Intentionally scratched glass at ambient conditions (SA.A, SA.H, SA.F).

Id	Thickness (mm)	σ_r (MPa)	Scratching device load (kg)	a_{s1} (μm)	a_{s2} (μm)	Effective load rate (N/s)	Effective stress rate (MPa/s)	t_f (s)	$Q(t_f)$ (kN)	$\sigma_E(t_f)$ (MPa)	$\sigma_n(t_f)$ (MPa)
SA.A-01	5.81	0	0.5	18.00	21.00	1297	21.1	1.76	2.287	37.5	37.5
SA.A-02	5.82	0	0.5	13.00	22.00	1257	20.5	1.41	1.770	29.0	29.0
SA.A-03	5.80	0	1.0	9.00	5.00	1315	21.3	2.08	2.734	44.7	44.7
SA.A-04	5.81	0	1.0	15.00	13.00	1330	21.5	2.41	3.207	52.4	52.4
SA.A-05	5.84	0	2.5	33.00	55.00	1292	21.0	1.87	2.412	39.5	39.5
SA.A-06	5.82	0	2.5	19.00	20.00	1302	21.2	1.67	2.176	35.7	35.7
SA.H-01	5.89	−77	0.5	18.00	21.00	1386	21.6	5.94	8.229	132.0	54.9
SA.H-02	5.88	−74	0.5	15.00	6.00	1378	21.8	4.34	5.978	96.7	22.8
SA.H-03	5.90	−77	1.0	32.00	13.00	1396	21.7	6.16	8.595	137.7	60.6
SA.H-04	5.87	−74	1.0	31.00	28.00	1391	21.9	4.84	6.738	108.7	34.8
SA.F-01	5.93	−112	0.5	20.00	21.00	1415	21.5	8.02	11.340	179.9	67.4
SA.F-02	5.91	−119	0.5	16.00	15.00	1402	21.6	7.00	9.806	156.4	37.5
SA.F-03	5.89	−129	1.0	25.00	20.00	1416	21.1	10.07	14.266	223.9	95.4
SA.F-04	5.89	−119	1.0	4.00	6.00	1413	21.2	9.21	13.006	205.1	86.2
SA.F-05	5.89	−125	2.5	52.00	38.00	1401	21.2	8.75	12.259	193.8	68.5
SA.F-06	5.92	−129	2.5	49.00	38.00	1394	21.5	7.02	9.794	156.2	27.7

C.3 Coaxial double ring test setup

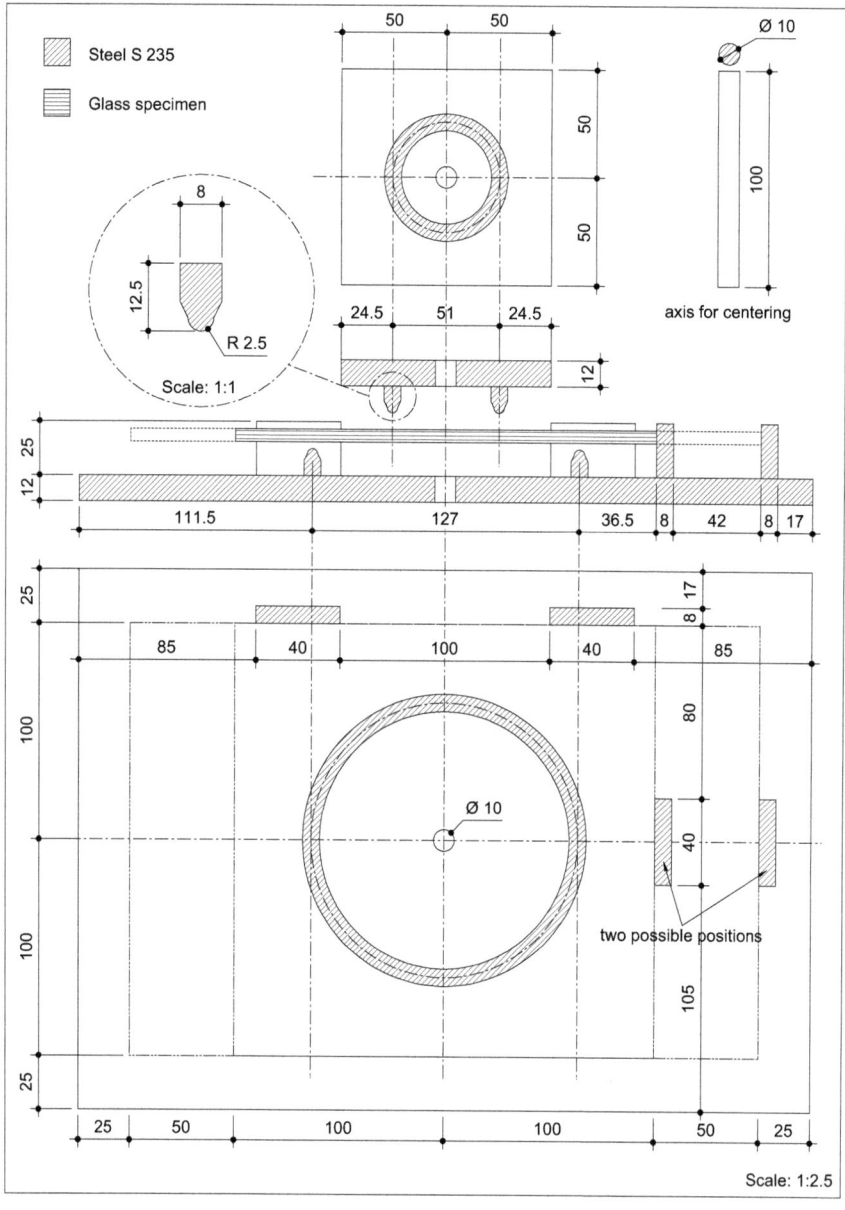

Figure C.6: Fabrication drawing for the coaxial double ring test equipment.

Appendix

D

Ontario Research Foundation Experimental Data

This appendix summarizes the experimental data from Johar (1981) (Phase II) and Johar (1982) (Phase III), which have been used in the present work. The following should be noted:

- Only specimens with the failure origin on the surface were used and are listed in the tables (no edge failures).
- The data are sorted by failure pressure.
- The letters A, B, and C in the specimen designations of Phase II represent three different glass manufacturers.
- All data were converted to SI units.

D.1 Phase II

Table D.1: Ontario Research Foundation data – 0.15 kPa/s.

Specimen	Thickness (mm)	Failure pressure (kPa)	Specimen	Thickness (mm)	Failure pressure (kPa)
C-06	5.78	4.00	A-03	5.73	5.52
B-01	5.92	4.48	B-08	5.89	5.72
B-07	5.89	4.48	C-10	5.78	5.86
C-07	5.79	4.62	C-08	5.79	6.07
C-05	5.80	4.69	A-10	5.69	6.21
C-09	5.78	4.69	B-02	5.88	6.21
C-11	5.78	4.83	B-04	5.89	6.41
C-02	5.80	4.96	A-07	5.72	6.89
A-04	5.73	5.17	A-06	5.72	7.24
C-04	5.80	5.24	A-09	5.69	7.24
C-01	5.80	5.31			

Table D.2: Ontario Research Foundation data – 1.5 kPa/s.

Specimen	Thickness (mm)	Failure pressure (kPa)	Specimen	Thickness (mm)	Failure pressure (kPa)
C-09	5.78	2.90	A-10	5.69	6.89
B-06	5.88	4.27	C-05	5.78	6.96
C-07	5.78	4.48	A-05	5.73	7.10
C-02	5.81	4.55	A-02	5.73	7.24
C-10	5.79	4.83	B-02	5.92	7.24
C-04	5.78	5.17	A-01	5.73	7.45
C-01	5.79	5.86	A-04	5.73	7.45
C-11	5.79	6.07	A-08	5.68	7.79
C-06	5.78	6.34	A-03	5.74	7.93
A-07	5.72	6.55	B-07	5.90	8.00
A-09	5.69	6.55	A-06	5.70	8.62
C-08	5.79	6.55			

Table D.3: Ontario Research Foundation data – 15 kPa/s.

Specimen	Thickness (mm)	Failure pressure (kPa)	Specimen	Thickness (mm)	Failure pressure (kPa)
C-08	5.79	4.48	B-09	5.89	6.96
B-05	5.88	4.83	A-10	5.68	7.03
C-05	5.79	5.03	A-09	5.69	7.10
C-09	5.78	5.38	B-06	5.86	7.24
A-04	5.72	5.65	C-02	5.80	7.38
C-06	5.79	5.79	A-06	5.73	7.58
A-03	5.73	6.00	B-10	5.89	7.58
A-05	5.73	6.07	B-07	5.89	7.79
C-01	5.80	6.76	B-08	5.89	8.14
C-03	5.81	6.89	A-02	5.73	8.96

D.2 Phase III

Table D.4: Ontario Research Foundation data – 0.0025 kPa/s.

Specimen	Thickness (mm)	Failure pressure (kPa)	Effective pressure rate (kPa/s)
18	5.82	2.34	0.0026
7	5.83	2.55	0.0025
8	5.82	2.83	0.0014
3	5.86	2.96	0.0024
1	5.78	3.10	0.0025
13	5.82	3.17	0.0028
19	5.80	3.24	0.0026
5	5.77	3.38	0.0024
10	5.84	3.45	0.0028
21	5.80	3.86	0.0023
2	5.83	3.93	0.0025
20	5.83	3.93	0.0026
17	5.80	4.21	0.0026
12	5.77	4.34	0.0027
4	5.83	4.69	0.0023
11	5.84	4.69	0.0023
16	5.83	4.83	0.0023

D.2. PHASE III

Table D.5: Ontario Research Foundation data – 0.025 kPa/s.

Specimen	Thickness (mm)	Failure pressure (kPa)	Effective pressure rate (kPa/s)
18	5.81	2.76	0.0248
16	5.80	3.17	0.0241
11	5.77	3.24	0.0269
15	5.81	3.45	0.0255
20	5.81	3.72	0.0255
8	5.82	3.93	0.0324
14	5.82	4.34	0.0255
7	5.83	4.41	0.0221
10	5.81	4.69	0.0241
21	5.81	4.69	0.0241
2	5.86	4.96	0.0248
19	5.80	5.10	0.0248
13	5.84	5.45	0.0248
17	5.82	6.21	0.0248

Table D.6: Ontario Research Foundation data – 0.25 kPa/s.

Specimen	Thickness (mm)	Failure pressure (kPa)	Effective pressure rate (kPa/s)
18	5.82	3.52	0.2482
11	5.82	4.76	0.2482
1	5.87	4.83	0.2344
15	5.82	4.83	0.2482
17	5.83	5.10	0.2482
22	5.80	5.38	0.2413
10	5.82	5.52	0.2551
2	5.90	5.79	0.2275
14	5.81	5.79	0.2482
20	5.84	5.93	0.2482
8	5.83	6.21	0.2413
7	5.79	6.34	0.2413
5	5.87	6.55	0.2344
3	5.87	6.62	0.2275
21	5.82	6.62	0.2413
19	5.83	6.89	0.2413

Table D.7: Ontario Research Foundation data – 2.5 kPa/s.

Specimen	Thickness (mm)	Failure pressure (kPa)	Effective pressure rate (kPa/s)
6	5.81	4.76	2.48
20	5.79	4.83	2.62
4	5.83	5.24	1.65
13	5.83	5.24	2.69
22	5.81	5.24	2.62
15	5.82	5.38	2.62
2	5.78	5.52	1.79
9	5.81	5.58	2.62
10	5.82	5.58	2.90
17	5.84	5.58	3.17
11	5.83	5.65	1.86
16	5.83	5.79	2.69
18	5.81	5.79	2.69
21	5.82	5.79	2.55
3	5.84	7.24	1.65
12	5.85	8.14	2.83
8	5.82	8.20	2.62
24	5.82	8.41	2.34
1	5.86	8.96	1.79

Table D.8: Ontario Research Foundation data – 25 kPa/s.

Specimen	Thickness (mm)	Failure pressure (kPa)	Effective pressure rate (kPa/s)
1	5.81	4.55	24.1
6	5.83	4.83	42.7
10	5.82	4.83	27.6
12	5.81	4.83	29.6
17	5.80	5.17	34.5
9	5.84	5.38	17.9
8	5.81	5.52	18.6
5	5.80	5.58	33.1
19	5.82	5.58	22.1
7	5.82	6.76	53.8
4	5.89	7.72	39.3
14	5.84	7.86	20.7
18	5.82	8.41	25.5
3	5.87	8.83	32.4
13	5.83	9.24	24.8

Appendix E

Statistical Fundamentals

This appendix provides some statistical fundamentals that are referred to in the text.

E.1 Statistical distribution functions

Table E.1: Continuous statistical distribution functions.

Type	PDF $f(x)$ CDF $F(x)$	Mean μ Variance σ^2
Normal	$f(x) = \dfrac{1}{\sigma\sqrt{2\pi}} \exp\left(-\dfrac{1}{2}\left(\dfrac{x-\mu}{\sigma}\right)^2\right)$	$\mu = \mu$
	$F(x) = \displaystyle\int_{-\infty}^{x} f(x)\,dx$	$\sigma^2 = \sigma^2$
Log-normal	$f(x) = \dfrac{1}{\zeta x \sqrt{2\pi}} \exp\left(-\dfrac{1}{2}\left(\dfrac{\ln x - \lambda}{\zeta}\right)^2\right)$	$\mu = \exp\left(\lambda + \dfrac{\zeta^2}{2}\right)$
	$F(x) = \displaystyle\int_{0}^{x} f(x)\,dx$	$\sigma^2 = \mu^2 \left(\exp(\zeta^2) - 1\right)$
Uniform	$f(x) = \dfrac{1}{b-a}$	$\mu = \dfrac{a+b}{2}$
	$F(x) = \dfrac{x-a}{b-a}$	$\sigma^2 = \dfrac{(b-a)^2}{12}$
Pareto	$f(x) = \dfrac{ab^a}{x^{a+1}}$	$\mu = \dfrac{ab}{a-1}$
	$F(x) = 1 - \left(\dfrac{b}{x}\right)^a$	$\sigma^2 = \dfrac{ab^2}{(a-1)^2(a-2)}$
Weibull	$f(x) = \dfrac{\beta}{\theta}\left(\dfrac{x}{\theta}\right)^{\beta-1} \cdot \exp\left(-\left(\dfrac{x}{\theta}\right)^\beta\right)$	$\mu = \theta \cdot \Gamma\left(1 + \dfrac{1}{\beta}\right)$
	$F(x) = 1 - \exp\left(-\left(\dfrac{x}{\theta}\right)^\beta\right)$	$\sigma^2 = \theta^2 \left[\Gamma\left(1 + \dfrac{2}{\beta}\right) - \Gamma^2\left(1 + \dfrac{1}{\beta}\right)\right]$

Table E.2: Discrete statistical distribution functions.

Type	PDF $f(x)$ CDF $F(x)$	Mean μ Variance σ^2	Notes
Poisson	$f(x) = \dfrac{e^{-\lambda}\lambda^x}{x!}$	$\mu = \lambda$	$x = 0, 1, 2, \ldots$
	$F(x) = \sum_{i=0}^{x} \dfrac{e^{-\lambda}\lambda^i}{i!}$	$\sigma^2 = \lambda$	

E.2 The empirical probability of failure

For some parameter estimation methods, for instance the least squares method (see Section E.3), an empirical probability distribution function for test data is required. In general, the probability density function (PDF) of the discrete random variable X is defined as

$$\hat{f}(x) = \begin{cases} p_i & \text{for } x = x_i \quad (i = 1, 2, 3, \ldots) \\ 0 & \text{for all other } x \end{cases} \qquad (E.1)$$

with p_i being the probability that the random variable X takes on the value x_i, which means

$$p_i = \frac{N_i}{N} \qquad (E.2)$$

in which N_i is the number of occurrences of the value i (generally 1 for test results) and N the total number of observations. The corresponding empirical cumulative distribution function is:

$$\hat{F}(x) = P(X \leq x) = \sum_{x_i \leq x} f(x_i) \qquad (E.3)$$

If test results are ordered such that i is the rank of the value x_i within all test results, the most obvious estimator is:

$$\hat{F}(x_i) = \frac{i}{N} \qquad (E.4)$$

While this estimator is very straightforward, it has at least two disadvantages. Firstly, the highest value cannot be represented on probability graphs and causes numerical problems. Secondly, it is very unlikely that the value with $\hat{F} = 1.0$ will be observed within relatively small samples. The largest value observed will thus lie below 1.0.

Values on the ordinate of a probability graph are actually random variables with a distribution of their own, which has a strong formal similarity to a beta distribution. The expectation value (mean rank) of this beta-distributed variable for the i-th value is

$$\hat{F}(x_i) = \frac{i}{N+1} \qquad (E.5)$$

and is independent of the observed values' distribution. The use of Equation (E.5) is recommended by many standard works on statistics. For large samples, the difference between $N + 1$ and N becomes very small. If the median (median rank) of the beta distribution is used instead of the expectation value, the estimator becomes[1]:

$$\hat{F}(x_i) = \frac{i - 0.3}{N + 0.4} \qquad (E.6)$$

There is no straightforward way of telling which estimator is more suitable. The difference for practical application is small. In order to ensure consistency with the European standard on the determination of the strength of glass EN 12603:2002, Equation (E.6) was used within the present work.

---

[1] This is a good approximation, the exact solution can only be found through the roots of a polynomial.

E.3 Parameter estimation, fitting and goodness-of-fit testing

E.3.1 Maximum likelihood method

The principle of the maximum likelihood method is that the parameters of the distribution function are fitted such that the probability (likelihood) of the observed random sample is maximized.

Let X be a random variable with the probability density function $f(x, \vec{\theta})$ where $\vec{\theta} = (\theta_1, \theta_2, \ldots, \theta_K)^T$ are the unknown constant parameters which need to be estimated. With the vector $\vec{x} = (\hat{x}_1, \hat{x}_2, \ldots, \hat{x}_N)$ containing the random samples from which the distribution parameters $\vec{\theta}$ are to be estimated, the likelihood function $L(\vec{x} \mid \vec{\theta})$ is given by the following product:

$$L(\vec{x} \mid \vec{\theta}) = \prod_{i=1}^{N} f(\hat{x}_i, \vec{\theta}) \tag{E.7}$$

The logarithmic likelihood function Λ, which is much easier to work with than L, is:

$$\Lambda(\vec{x} \mid \vec{\theta}) = \ln L(\vec{x} \mid \vec{\theta}) = \sum_{i=1}^{N} \ln f(\hat{x}_i, \vec{\theta}) \tag{E.8}$$

The maximum likelihood estimators of the parameters $\theta_1, \theta_2, \ldots, \theta_K$ are obtained by solving the following optimization problem:

$$\min_{\vec{\theta}}(-L(\vec{x} \mid \vec{\theta})) \quad \text{or} \quad \min_{\vec{\theta}}(-\Lambda(\vec{x} \mid \vec{\theta})) \tag{E.9}$$

As can be seen from the equations, the maximum likelihood method is independent of any kind of ranks or plotting methods (cf. Section E.2). The maximum likelihood estimators have a higher probability of being close to the quantities to be estimated than the point estimators obtained with the method of moments have. (Faber, 2003; www.weibull.com)

E.3.2 Least squares method

To obtain the coefficient estimates, the least squares method minimizes the summed square of residuals. The residual for the i-th data point Δ_i is defined as the difference between the observed response value y_i and the fitted response value $\hat{y}_i$, and is identified as the error associated with the data. The summed square of the residuals (error estimate) is given by:

$$S = \sum_{i=1}^{I} \Delta_i^2 = \sum_{i=1}^{I} (y_i - \hat{y}_i)^2 \tag{E.10}$$

in which I is the number of data points included in the fit. (Matlab Doc 2005)

E.3.3 Method of moments

EN 12603:2002, the European standard for the analysis of glass strength data, is based on the method of moments. For uncensored samples, the following Weibull parameter point estimates are given:

$$\hat{\theta} = \exp\left[\frac{1}{N}\sum_{i=1}^{N} \ln x_i + 0.5772 \frac{1}{\hat{\beta}}\right] \quad ; \quad \hat{\beta} = N\kappa_N \cdot \left(\frac{s}{N-s}\sum_{i=s+1}^{N} \ln x_i - \sum_{i=1}^{s} \ln x_i\right)^{-1} \tag{E.11}$$

$\hat{\theta}$ and $\hat{\beta}$ are the point estimates for the shape and scale parameters respectively. N is the sample size, x_i is the i-th sample, s is the largest integer $< 0.84N$. The factor κ_N is a function of N and is provided in a table (examples: $N = 5 \Rightarrow \kappa_N = 1.2674$, $N = 10 \Rightarrow \kappa_N = 1.3644$, $N = 20 \Rightarrow \kappa_N = 1.4192$).

While the maximum likelihood method and the least squares method can be used to estimate parameters of any model, the point estimators in Equation (E.11) can *only* be used to estimate the parameters of a two-parameter Weibull distribution. Their main advantage is their simplicity.

E.3.4 Modified Anderson-Darling goodness-of-fit test

The Anderson-Darling test (Stephens, 1974) is used to test whether a sample of data comes from a population with a specific statistical distribution. It is particularly suitable for the assessment of the goodness-of-fit of glass strength data for the following reasons:

- The older and widely used Kolmogorov-Smirnov test tends to be more sensitive near the center of the distribution than at the tails. The Anderson-Darling test gives more weight to the tails.
- The K-S test is no longer valid if the distribution parameters are estimated from the data. This is a very severe limitation that the A-D test does not suffer from.
- The A-D test was found to be more sensitive to lack of fit of a two-parameter Weibull distribution than the Kolmogorov-Smirnov and the Shapiro-Wilk test (another well known goodness-of-fit test) (Evans et al., 1989).

In contrast to the K-S test, the A-D test makes use of distribution-specific critical values (see p-value below).

With the vector $\vec{x} = (\hat{x}_1, \hat{x}_2, \ldots, \hat{x}_N)$ containing the samples (measured values) in ascending order, the Anderson-Darling goodness-of-fit test is performed as follows:

- The null hypothesis is that the data in $\vec{x}$ come from a Weibull-distributed population.
- The Weibull parameters are estimated using one of the methods described earlier in the present section.
- The value of the Weibull cumulative distribution function (see Section E.1) is calculated for all samples, yielding the new vector $\vec{z}_{(i)} = F_{\text{Wb}}(\vec{x}_{(i)})$.
- The Anderson-Darling statistic A^2 is calculated as follows (Stephens, 1974):

$$A^2 = -\frac{\sum_{i=1}^{N}(2i-1)\left[\ln \vec{z}_{(i)} + \ln\left(1 - \vec{z}_{(N+1-i)}\right)\right]}{N} - N \qquad (E.12)$$

- The observed significance level probability (p-value) is (START 2003:5):

$$p_{\text{AD}} = \frac{1}{1 + \exp\left[-0.1 + 1.24 \cdot \ln(\check{A}) + 4.48 \cdot \check{A}\right]} \quad \text{with} \quad \check{A} = (1 + \frac{0.2}{\sqrt{N}}) \cdot A^2 \qquad (E.13)$$

- If $p_{\text{AD}} < 0.05$, then the Weibull assumption is rejected and the error committed is less than 5%.

The Anderson-Darling test is not available in common software packages. The following function for Matlab (Matlab 2005) has therefore been written and used for the present research project:

```
function res = AndersonDarlingWeibull( samples, wbshape, wbscale )
    X = sort(samples);              % samples in ascending order
    Z = wblcdf(X, wbscale, wbshape); % CDF of the Weibull dist. for each value in X
    n = length(X);                  % number of samples

    s = 0;
    for i=1:n
        s = s + (2*i - 1) * ( log(Z(i)) + log(1 - Z(n+1-i)) );
    end

    AD = - s / n - n
    ADbreve = (1 + 0.2/sqrt(n))*AD
    OSL = 1 / ( 1 + exp(-0.1 + 1.24*log(ADbreve) + 4.48*(ADbreve)))

    res = [AD, ADbreve, OSL]
end
```

Appendix

F

Predictive Modelling Example: EPFL-ICOM Column Buckling Tests

As an example, this appendix shows the application of GlassFunctionsForExcel for predicting the results of column buckling laboratory tests[1]. Such tests were performed by *Luible* at EPFL-ICOM (Luible, 2004a). He tested a wide range of specimen geometries, so that only a few specimens were identical. The largest series of identical geometry, which comprises four specimens, is used for the present example. The relevant experimental data are summarized in Table F.1.

Table F.1:
Experimental data of column buckling tests by *Luible* (Luible, 2004a) on 800 mm × 200 mm heat strengthened glass specimens.

Id	Imperfection at midspan (mm)	Residual surface stress (MPa)	Buckling stress at failure (MPa)
FL-TVG10/800-1	1.04	−50.2	8.45
FL-TVG10/800-2	1.55	−45.8	8.14
FL-TVG10/800-3	1.66	−46.9	8.11
FL-TVG10/800-4	1.55	−48.0	8.13
Average	1.45	−47.7	8.21

The 800 mm high, 200 mm wide and 10 mm thick heat strengthened glass specimens were tested in a standard column buckling test setup, i. e. the displacement perpendicular to the specimen was restrained along the short edges and rotation remained unconstrained. An imposed deformation of 2.5 μm/s was applied to one short edge.

The results of these experiments were predicted using the process outlined in Section 8.2:

- **Material model.** The parameters were chosen according to the findings discussed in Chapter 6, i. e. $K_{Ic} = 0.75$ MPa m$^{0.5}$, $v_0 = 10$ μm/s, $n = 16$, $Y = 1.12$, $\theta_0 = 63$ MPa, $m_0 = 8.1$. Furthermore, $\sigma_r = -47.7$ MPa (average value from Table F.1), $E = 74$ GPa and $v = 0.23$ were used.

- **Structural model.** The glass specimen's mechanical behaviour was analysed with the finite element software package Abaqus (ABAQUS 2004). The specimen was represented by 120 8-node shell elements with quadratic interpolation (S8R). The analysis involved the following steps (Figure F.2):

  ◇ Determination of the first eigenvector of the specimen.

  ◇ Definition of the initial geometry as the first eigenvector, scaled to the average measured initial deformation from Table F.1.

  ◇ Non-linear finite element analysis.

---
[1] This simple case could also be solved analytically, without a finite element model. This is not done here, since the idea of the present section is to show the general process that can be used for arbitrary geometry and loading conditions.

⋄ Generation of output files containing (a) major and minor in-plane principal stresses in all finite elements as a function of the imposed deformation, (b) surface areas of all finite elements, and (c) reaction forces[2] as a function of the imposed deformation.

Model generation, analysis and result extraction were again performed by the scripts already mentioned in Section 6.5 that were developed by the author to automate these tasks.

- **Action model.** The imposed deformation was applied to one short edge at a constant rate of 2.5 µm/s. GlassFunctionsForExcel offers a specific function (see below) for the simple but frequent case of constant action rate[3], which handles action history generation automatically.
- **Lifetime prediction model.** GlassFunctionsForExcel's RSFPMProbFailureCARGeneral function calculates the failure probability for constant action rate loading and is thus suitable for the task at hand. The function takes as input a material model, a structural model, the maximum action and the action rate. In order to use it, the following needs to be typed into a cell of an Excel worksheet (cf. Section 8.5.2 and Section H.2): =RSFPMProbFailureCARGeneral(⟨maximum action⟩, ⟨action rate⟩, ⟨number of action history steps to generate⟩, ⟨|σ_r|⟩, ⟨θ_0⟩, ⟨m_0⟩, ⟨Y⟩, ⟨K_{Ic}⟩, ⟨v_0⟩, ⟨n⟩, ⟨path to the file containing σ_1 and A_i⟩, ⟨path to the file containing σ_2 and A_i (if empty, an equibiaxial stress field is assumed)⟩, ⟨segments to use when accounting for biaxial stress fields⟩).

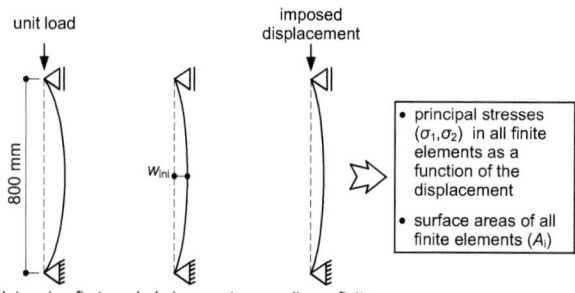

Figure F.2:
Structural analysis of column buckling experiments.

In Figure F.3, the predicted failure probabilities are compared to the measured ones.[4] It is seen that the error is less than 10%, which can be considered acceptable.

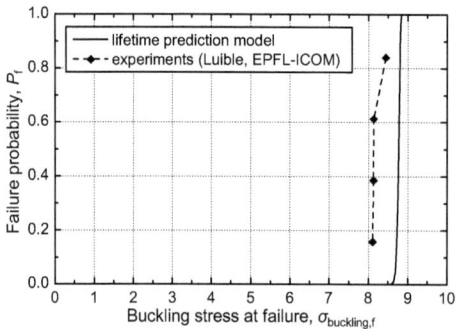

Figure F.3:
Comparison of the predicted and measured failure probability as a function of the buckling stress (= reaction force divided by the sectional area).

---

[2] The reaction forces are not required for the lifetime prediction model, but for comparison with the experimental data.
[3] E. g. constant stress rate, constant load rate or constant deformation rate.
[4] The estimator for the empirical probability that is used to plot the experimental data is discussed in Section E.2.

Appendix G
Source Code Listings

This appendix contains source code fragments that are referred to in the text.

G.1 Functions for the analysis of large-scale experiments

In order to provide the reader with an idea of the Matlab implementation, the source code of some *key functions* is listed below. The reproduction of all code required for the simulations presented in this study would take up far too much space and would be of limited interest to most readers. Those who are interested in the complete code may contact the author by e-mail (matthias@haldimann.info).

G.1.1 Failure probability, general formulation (pf_general.m)

```matlab
function out = pf_general(acthist, sigmaresabs, ...
    theta_0, m_0, Y, K_Ic, v_0, n, FEAareas, FEAactions, FEAs1, FEAs2, numofseg)
% Implements the generalized fracture mechanics model for glass elements.
% (random surface flaw population).
%   Returns the probability of failure using the general formulation.

%% constants, input check, residual surface stress ##################################
A_0 = 1.0;
% the constant 1 / U / theta_0^(n-2)
% Note: don't calculate using U, this causes division by 0 error for v_0=0
constst = (n-2) * v_0 * (Y*sqrt(pi))^2 / (2 * K_Ic^2) / (theta_0^(n-2));

% check if FEA data covers all action intensities of the action history
if max(acthist(:,2)) > max(FEAactions)
    error('FEA data must cover all action intensities in the history.')
end

% take residual surface compression stress into account
FEAs1 = FEAs1 - sigmaresabs;
FEAs2 = FEAs2 - sigmaresabs;

%% find stress history from action history ##########################################
% (columns => actionsteps, rows => elements)
stresshist1 = zeros(max(size(FEAareas)), size(acthist,1));
stresshist2 = zeros(max(size(FEAareas)), size(acthist,1));

for s=1:size(acthist,1)
```

```
        loindex = 0; upindex = 0; bias = 0;
        loadIntensity = acthist(s,2);

        % find indices of closest values and bias
        for l = 1:length(FEAactions)
            if(FEAactions(l) >= loadIntensity)
                loindex = l-1;       % index of the lower neighbour
                upindex = l;         % index of the upper neighbour

                if (loindex < 1)     % special case if loadIntensity=loadArray[1]
                    loindex = 1; bias = 0;
                else                 % if not => normal case
                    bias = ( loadIntensity - FEAactions(loindex) ) / ...
                        ( FEAactions(upindex) - FEAactions(loindex) ) ;
                end
                break;
            end
        end

        % loindex, upindex and bias are valid for both sigma_1 and sigma_2
        stresshist1(:,s)= FEAs1(:,loindex) + ...
            bias * (FEAs1(:,upindex) - FEAs1(:,loindex));

        if numofseg > 0
            stresshist2(:,s)= FEAs2(:,loindex) + ...
                bias * (FEAs2(:,upindex) - FEAs2(:,loindex));
        end
end
clear FEAs1; clear FEAs2; clear FEAactions;

%% sum over timesteps, segments and finite elements if numofseg > 0 ################
if numofseg > 0
    % prepare segmenting data
    dphi = 0.5 * pi / numofseg;           % segment width
    phi = zeros(numofseg, 1, 'double');   % orientations (middle of segments)
    for k = 1:numofseg
        phi(k) = 0.5 * dphi + (k-1) * dphi;
    end

    % do the triple sum
    sumarea = 0;
    for i = 1:size(FEAareas)
        sumangle = 0;
        for k = 1:size(phi)
            sumtime = 0;
            maxfn = 0;
            for j=1:size(stresshist1,2)
                % calculate the crack opening stress
                % (note: inline! function call takes 80 times longer...)
                sn = stresshist1(i,j) * cos(phi(k))^2 + ...
                    stresshist2(i,j) * sin(phi(k))^2;
                if sn > 0    % only consider traction stresses
                    sumtime = sumtime + (sn^n * acthist(j,1));
                    % acthist(j,1) = duration of time increment
                    % note: sn^n is very costly in terms of CPU time
                    maxfn_tmp = (sn / theta_0)^(n-2) + conststs * sumtime;
                    if maxfn_tmp > maxfn    % is this the max?
                        maxfn = maxfn_tmp;
```

G.1. FUNCTIONS FOR THE ANALYSIS OF LARGE-SCALE EXPERIMENTS

```
                                end
                            end
                        end
                        sumangle = sumangle + dphi * (maxfn^(m_0/(n-2)));
                    end
                    sumarea = sumarea +  sumangle * FEAareas(i);
                end
            %% sum over timesteps and finite elements only if numofseg == 0 ####################
            else
                sumarea = 0;
                for i = 1:size(FEAareas)
                    sumtime = 0;
                    maxfn = 0;
                    for j=1:size(stresshist1,2)
                        sn = stresshist1(i,j);
                        if sn > 0    % only consider traction stresses; sigma_1
                            sumtime = sumtime + (sn^n * acthist(j,1));
                            % acthist(j,1) = duration of time increment
                            % note: sn^n is very costly in terms of CPU time
                            maxfn_tmp = (sn / theta_0)^(n-2) + constst * sumtime;
                            if maxfn_tmp > maxfn
                                maxfn = maxfn_tmp;       % is this the max?
                            end
                        end
                    end
                    sumangle = (pi / 2) * (maxfn)^(m_0/(n-2));
                    sumarea = sumarea + sumangle * FEAareas(i);
                end
            end

            %% return failure probability ###########################################
            out = 1 - exp(-(1/A_0) * (2/pi) * sumarea);

        end
```

G.1.2 Parameter fitting (paramFit.m)

```
function [optimparams, fval, exitflag] = ...
    paramFit(paramvect, gp, gd, fitmethod, paramtype)
% General glass model paramter fitting.
%
% SUPPORTED FITTING ALGORITHMS:
%   - log maximum likelihood method or
%   - least squares method (horizontal and vertical)
%
% RANGE OF VALIDITY:
% The action rate (= load-, stress or displacement rate depending on the
% FEAdata provided) must be constant.
%
% PARAMETERS:
% paramtype:
%    string indicating the parameters to find
%    'surfcond' => paramvect = [theta_0 m_0]
%    'v_0'      => paramvect = [v_0]
%
```

```
% paramvect:
%    the guess values to start with
%
% gp, gd:
%    glass properties and glass data structs; see createParamStruct.m for details
%
% fitmethod:
%    string indicating the fitting method to use ('maxlik', 'lsqvert' or 'lsqhor')
%
% NOTE:
%    This function is very performance critical. It is, therfore, unavoidable to
%    cut back on clarity and elegance of the code in order to improve performance.

% optimizer options
options = optimset;
options.TolX = 1e-5;
options.MaxIter = 500;
options.MaxFunEvals = 2000;
disp(['TolX: ' num2str(options.TolX) ' MaxIter: ' num2str(options.MaxIter)]);

switch paramtype
    case 'surfcond' % surface condition parameter fit
        % Convert theta_0 Pa -> MPa to avoid bad minimizing
        % behaviour due to uneven parameter vector scaling
        paramvect(1) = paramvect(1) / 1e6;

        switch fitmethod
            case 'maxlik'
                % find min of the minus-log-likelihood function
                [optimparams,fval,exitflag] = fmincon(@loglhfun_surfcond, paramvect,...
                    [], [], [], [], [1e-10 1e-10], [200 200], [], options)
            case {'lsqvert', 'lsqhor'}
                if (strcmp(fitmethod, 'lsqvert'))
                    fh = @vError_surfcond;
                else
                    fh = @Error_surfcond;
                end
                % minimize the delta-squares of the p_f (vertical)
                [optimparams,resnorm,residual,exitflag,output] = ...
                    lsqnonlin(fh, paramvect, [1e-10 1e-10], [200 200], options);
        end
        optimparams(1) = optimparams(1) * 1e6;  % scale back theta_0

    case 'v_0'  % v_0 only fit
        paramvect(1) = paramvect(1) * 1e3;

        switch fitmethod
            case 'maxlik'
                [optimparams,fval,exitflag] = fmincon(@loglhfun_v_0, paramvect, ...
                    [], [], [], [], [1e-10], [1e10], [], options)
            case {'lsqvert', 'lsqhor'}
                if (strcmp(fitmethod, 'lsqvert'))
                    fh = @vError_v_0;
                else
                    fh = @Error_v_0;
                end
                % minimize the delta-squares
                [optimparams,resnorm,residual,exitflag,output] = ...
```

G.1. FUNCTIONS FOR THE ANALYSIS OF LARGE-SCALE EXPERIMENTS

```
                    lsqnonlin(fh, paramvect, [1e-10], [1e10], options);
        end
        optimparams(1) = optimparams(1) / 1e3;   % scale back v_0
end

if (strcmp(fitmethod, 'lsqvert') | strcmp(fitmethod, 'lsqhor'))
    fval = sum(residual .^ 2);   % set fval for least-square fitting types
end

%% NESTED FUNCTIONS ################################################
    %% functions for max likelihood fit ---------------------------------
    function out = loglhfun_surfcond(pv)
        % update the properties struct with the params to optimize
        gp.theta_0 = pv(1) * 1e6;   % scaling back MPa -> Pa
        gp.m_0 = pv(2);

        sum = 0;
        for edcount=1:size(gd.EXPdata,1)
            failureload = gd.EXPdata(edcount,1);
            sum = sum + log(...
                pdf_CAR(failureload, gd.actionrate, gp.sigmaresabs, ...
                    gp.theta_0, gp.m_0, gp.Y, gp.K_Ic, gp.v_0, gp.n, ...
                    gd.FEAareas, gd.FEAactions, gd.FEAs1, gd.FEAs2, ...
                    gp.numofseg));
        end
        out = -sum;   % - => maximum becomes minimum
        disp([gp.theta_0 gp.m_0 gp.v_0 gp.n out]);   % diagnostic output
    end

    function out = loglhfun_v_0(pv)
        % update the properties struct with the params to optimize
        gp.v_0 = pv(1) / 1e3;   % scaling back mm -> m

        sum = 0;
        for edcount=1:size(gd.EXPdata,1)
            failureload = gd.EXPdata(edcount,1);
            sum = sum + log(...
                pdf_CAR(failureload, gd.actionrate, gp.sigmaresabs, ...
                    gp.theta_0, gp.m_0, gp.Y, gp.K_Ic, gp.v_0, gp.n, ...
                    gd.FEAareas, gd.FEAactions, gd.FEAs1, gd.FEAs2, gp.numofseg));
        end
        out = -sum;   % - => maximum becomes minimum
        disp([gp.theta_0 gp.m_0 gp.v_0 gp.n out]);   % diagnostic output
    end

    %% functions for p_f (vertical) least-squares fit -------------------
    function [errvect] = vError_surfcond(pv)
        % update the properties struct with the params to optimize
        gp.theta_0 = pv(1) * 1e6;   % scaling back MPa -> Pa
        gp.m_0 = pv(2);

        % get the vector of the deltas, scale for better numeric behaviour
        [errvect] = modelErrorCARVertical(gp, gd) * 1000;
    end

    function [errvect] = vError_v_0(pv)
        % update the properties struct with the params to optimize
        gp.v_0 = pv(1) / 1e3;   % scaling back mm -> m
```

```
            % get the vector of the deltas, scale for better numeric behaviour
            [errvect]  = modelErrorCARVertical(gp, gd) * 1000;
        end

        %% functions for action (horizontal) least-squares fit ------------------------
        function [errvect] = Error_surfcond(pv)
            % update the properties struct with the params to optimize
            gp.theta_0 = pv(1) * 1e6;    % scaling back MPa -> Pa
            gp.m_0 = pv(2);

            % get the vector of the deltas, scale for better numeric behaviour
            [errvect]  = modelErrorCAR(gp, gd) * 1000;
        end

        function [errvect] = Error_v_0(pv)
            % update the properties struct with the params to optimize
            gp.v_0 = pv(1) / 1e3;    % scaling back mm -> m

            % get the vector of the deltas, scale for better numeric behaviour
            [errvect]  = modelErrorCAR(gp, gd) * 1000;
        end
end
```

G.1.3 Model error vector using Δaction (modelErrorCAR.m)

```
function [errvect] = modelErrorCAR(gp, gd)
% The model error vector (delta-action at all P_f in experimental data).
%
% PARAMETERS:
%    gp, gd: glass parameter and glass data structs

disp([gp.theta_0 gp.m_0 gp.v_0 gp.n]);   % informative output

%% get the model data ##############################################
% IMPORTANT: the actionrange must cover all actions that are required
% to attain p_f = 1 for all parameter sets used by the optimizer or user!
mdata = zeros(length(gd.FEAactions),2);
lastpf = 0;   % used to avoid multiple calculations for pf=1
for i=1:size(mdata,1)
    failureload = gd.FEAactions(i);
    acthist = generateActionCLR(failureload, gd.actionrate, 50);
    % get probability of failure
    mdata(i,1) = failureload;
    if lastpf < 1   % if last P_f was not yet 1.0, calculate P_f
        mdata(i,2) = pf_general(acthist, gp.sigmaresabs, ...
                    gp.theta_0, gp.m_0, gp.Y, gp.K_Ic, gp.v_0, gp.n, ...
                    gd.FEAareas, gd.FEAactions, gd.FEAs1, gd.FEAs2, ...
                    gp.numofseg);
    else   % if last P_f was already 1.0, current P_f is also 1.0
        mdata(i,2) = 1.0;   % avoids useless calculation => speed
    end
    lastpf = mdata(i,2);
end

%% remove duplicate 0 and 0 P_f values from mdata (required for P_f interpolation) ##
```

G.1. FUNCTIONS FOR THE ANALYSIS OF LARGE-SCALE EXPERIMENTS

```
    for i=1:size(mdata,1)       % remove 0-value duplicates at the beginning
        if mdata(i,2) > 0
            if i > 2
                mdata(1:i-2,:) = [];
                break;
            end
        end
    end

40  for i=size(mdata,1):-1:1    % remove 1-value duplicates at the end
        if mdata(i,2) < 1
            if i < (size(mdata,1)-1)
                mdata(i+2:size(mdata,1),:) = [];
                break;
            end
        end
    end

    %% interpolate model data at the experimental P_f values ############################
50  mdata2 = zeros(size(gd.EXPdata));
    mdata2(:,2) = gd.EXPdata(:,2);
    mdata2(:,1) = interp1(mdata(:,2),mdata(:,1),mdata2(:,2));

    errvect = gd.EXPdata(:,1) - mdata2(:,1);
    end
```

G.1.4 Model error vector using ΔP_f (modelErrorCARVertical.m)

```
1   function [errvect] = modelErrorCARVertical(gp, gd)
    % The VERTICAL model error (delta p_f at all failureactions in exp. data.)
    %
    % PARAMETERS:
    %   gp, gd: glass parameter and glass data structs

    disp([gp.theta_0 gp.m_0 gp.v_0 gp.n]);    % informative output

    locations = gd.EXPdata(:,1);    % locations at which to get model data

10  mdata = zeros(length(locations),2);
    for i=1:length(locations)
        failureload = locations(i);
        acthist = generateActionCLR(failureload, gd.actionrate, 50);
        mdata(i,1) = failureload;
        mdata(i,2) = pf_general(acthist, gp.sigmaresabs, ...
            gp.theta_0, gp.m_0, gp.Y, gp.K_Ic, gp.v_0, gp.n, ...
            gd.FEAareas, gd.FEAactions, gd.FEAs1, gd.FEAs2, ...
            gp.numofseg);
20  end
    errvect = gd.EXPdata(:,2) - mdata(:,2);
    end
```

G.2 Programming language performance comparison

G.2.1 Benchmarking algorithm (C++ version)

```cpp
#include <iostream>
#include <math.h>
#include <time.h>
using namespace std;

int main()
{
        const int measurements = 1000;
        const int arrsize = (int)1e5;
        const double x = 100;
        const double n = 16;
        double sumofarr = 0;

        clock_t start = clock();
        for (int k = 0; k < measurements; k++)
        {
                double *arrin = new double[arrsize];
                double *arrout = new double[arrsize];

                for (int i = 0; i < arrsize; i++)
                {
                        arrin[i] = (double)i * (x / (double)arrsize);
                }

                for (int i = 0; i < arrsize; i++)
                {
                        arrout[i] = pow(arrin[i], n);
                }

                sumofarr = 0;
                for (int i = 0; i < arrsize; i++)
                {
                        sumofarr += arrout[i];
                }
        }
        double time = ((double)clock() - start) / CLOCKS_PER_SEC / (double)measurements;

        printf("C++ speed test. Average over %d measurements.\n", measurements);
        printf("Average time to complete: %lf\n", time);
        printf("Sum of array: %e (Size: %d)\n", sumofarr, arrsize);
        return 0;
}
```

Appendix

H Documentation for the 'GlassTools' Software

The following documentation for the *GlassTools* software has been shortened. It is intended for readers who are interested in the concept and the possibilities of the software and does not include all classes required for *GlassTools* to work. Those who would like to use the software should refer to the electronic documentation (HTML) that comes with the source code. It is not only complete but also much more convenient because it is structured, cross-linked and includes an index.

Readers who are interested in the source code and/or the electronic documentation for *GlassTools* may contact the author by e-mail (matthias@haldimann.info).

H.1 Class Hierarchy

MH::GlassFunctionsForExcel 180

MH::SSFM 183

MH::RSFPM 185

MH::GlassElement 187

MH::ActionObj 188

MH::ActionObj::Action 189

MH::ActionObj::ActionCollection 189

MH::MaterialObj 190

MH::ProbabilityDistribution 191

    MH::NormalDist 193

    MH::LogNormalDist 193

    MH::UniformDist 194

    MH::WeibullDist 194

    MH::DeterministicVariable 195

H.2 MH::GlassFunctionsForExcel Class Reference

Detailed Description

This is an Excel automation add-in that offers various functions for the analysis and design of structural glass elements.
Note two important things:

- Static functions cannot be called by Excel.
- If functions should be able to display exception messages in Excel, they must return an object (not a double).

Public Member Functions

- object **ConvertGFPMParams_GetScale** (double m_GFPM, double k_GFPM, double Y, double K_Ic, double v_0, double n)

    Find θ_0 from the GFPM surface flaw parameters m and k.

- object **ConvertGFPMParams_GetShape** (double m_GFPM, double n)

    Find m_0 from the GFPM surface flaw parameter m.

- object **ConvertNonInertRORResults_Scale** (double testscaleparam, double testshapeparam, double teststresssurface, double teststressrate, double Y, double K_Ic, double v_0, double n)

    Find θ_0 from ambient test data.

- object **ConvertNonInertRORResults_Shape** (double testshapeparam, double n)

    Find m_0 from ambient test data.

- double **CrackDepthAfterLoadingExact** (double sigma_eq_1s, double a_i, double Y, double K_Ic, double v_0, double n)

    Returns the crack depth after loading for some 1s-equivalent static stress. If the crack fails, -1 is returned.

- double **CriticalCrackDepth** (double InertStrength, double Y, double K_Ic)

    Critical crack depth (inert strength of a crack).

- object **DecompressedSurfaceArea** (double actionIntensity, double sigmaresabs, string feadatafilepath)

    Returns the decompressed surface area of an element (of both faces if the FEA data contains them).

- object **EqOneSecStressEquibiaxCAR** (double maxaction, double actionrate, double sigmaresabs, double n, string feadatafilepath)

    Returns the one-second equivalent stress for an element with a biaxial stress field and a constant action rate. (The FEA datafile must contain the action-stress relationship for ONE element.).

- object **EqOneSecStressOnRefSurface** (double actionDuration, double actionIntensity, double sigmaresabs, double theta_0, double m_0, double Y, double K_Ic, double v_0, double n, string feadatafilepath1, string feadatafilepath2)

    Returns the equivalent one-second uniform first principal stress on the reference surface ($\bar{\sigma}$) for a constant action.

- double **EquivalentStaticStressCSR** (double sigmamax, double n)

    Equivalent static stress for a period of constant stress rate (relating to the same duration as the original stress history).

- double **EquivalentTRefStressCS** (double stress, double duration, double n, double t_ref)

    t_{ref}-equivalent static stress for a period of constant stress

- double **EquivalentTRefStressCSR** (double stressrate, double sigmamax, double n, double tref)

    t_{ref}-equivalent static stress for a period of constant stress rate

- object **EquivArea** (double actionIntensity, double sigmaresabs, double m_0, double n, string feadatafilepath)

    Equivalent area of an element for a given action intensity.

- double **FailureStressCSApprox** (double timetofailure, double a_i, double Y, double K_Ic, double v_0, double n)

    Approximate formulation of FailureStressCSExact; assuming $a_{final} \gg a_i$ (enables direct solution).

- double **FailureStressCSExact** (double timetofailure, double a_i, double Y, double K_Ic, double v_0, double n)

    Failure stress for constant stress loading (exact formulation).

H.2. MH::GLASSFUNCTIONSFOREXCEL CLASS REFERENCE

- double **FailureStressCSRApprox** (double stressrate, double a_i, double Y, double K_Ic, double v_0, double n)

  Approximate formulation of FailureStressCSRExact; assuming $a_{final} \gg a_i$ (enables direct solution).

- double **FailureStressCSRExact** (double stressrate, double a_i, double Y, double K_Ic, double v_0, double n)

  Failure stress for constant stress rate loading (exact formulation).

- string **GetInfo** ()

  Returns a version information message.

- double **InertStrengthSingleCrack** (double a, double Y, double K_Ic)

  Inert strength of a single crack.

- double **InitialCrackSizeMonIncreasingExact** (double sigma_eq_1s, double failurestress, double Y, double K_Ic, double v_0, double n)

  Initial crack size that leads to failure after the MONOTONOUSLY INCREASING (or constant) stress history characterized by the one-second equivalent static stress and the failure stress (exact formulation).;

- object **InterExtrapolateLinear** (double x, Excel.Range xrange, Excel.Range yrange)

  Linear interpolation AND extrapolation using Excel cell ranges.

- object **InterpolateLinear** (double x, Excel.Range xrange, Excel.Range yrange)

  Linear interpolation using Excel cell ranges.

- double **k_bar** (double theta_0, double m_0, double Y, double K_Ic, double v_0, double n)

  Combined parameter $\bar{k}$.

- double **LogNormInv** (double probability, double mean, double stdev)

  Returns the inverse cumulative distribution function for a Log-normal distribution (missing in Excel).

- double **m_bar** (double m_0, double n)

  Combined parameter $\bar{m}$.

- object **MaxPrincStressInElement** (double action, double sigmaresabs, string feadatafilepath)

  Returns the maximum first principal stress in an element at a given action level.

- object **MaxPrincStressRateInElement** (double actionIntensity, double actionRate, double sigmaresabs, string feadatafilepath)

  Returns the maximum principal stress rate in an element at a given action level and action rate.

- double **PfEquibiaxCSR** (double sigma_failure, double stressrate, double theta_0, double m_0, double Y, double K_Ic, double v_0, double n, double A)

  Ambient strength of a surface of area A exposed to a homogenous equibiaxial stress field at a constant stress rate (analytical formulation).

- double **PfEquibiaxFromOneSecEq** (double sigma_1s, double theta_0, double m_0, double Y, double K_Ic, double v_0, double n, double A)

  Calculates the failure probability from the one-second equivalent stress for an equibiaxial stress field.

- double **PfEquibiaxInert** (double sigma_failure, double theta_0, double m_0, double A)

  Inert strength of a surface of area A exposed to a homogenous equibiaxial stress field.

- object **ProbabilityOfFailureSimplif** (double sigma_bar, double sigmaresabs, double theta_0, double m_0, double Y, double K_Ic, double v_0, double n)

  Returns the failure probability for a given $\bar{\sigma}$ (simplified formulation).

- double **ReferenceAmbientStrength** (double P_f, double theta_0, double m_0, double Y, double K_Ic, double v_0, double n)

  Reference ambient strength f_0.

- double **ReferenceInertStrength** (double P_f, double theta_0, double m_0)

  Reference inert strength $f_{0,inert}$.

- object **RSFPMConstantActionFnPf** (double pfAtEndOfLifetime, int requiredLifetime, double sigmaresabs, double theta_0, double m_0, double Y, double K_Ic, double v_0, double n, string feadatafilepath)

  Finds the constant action that yields the given failure probability when applied constantly during a given lifetime.

- object **RSFPMProbFailureCARGeneral** (double maxaction, double actionrate, int numofsteps, double sigmaresabs, double theta_0, double m_0, double Y, double K_Ic, double v_0, double n, string feadatafilepath1, string feadatafilepath2, int numofseg)
  Failure probability for constant action rate loading; General formulation.

- object **RSFPMProbFailureInertCAR** (double maxaction, int numofsteps, double sigmaresabs, double theta_0, double m_0, double Y, double K_Ic, string feadatafilepath1, string feadatafilepath2, int numofseg)
  Inert failure probability for constant action rate loading.

- object **SurfaceArea** (string feadatafilepath)
  Returns the total surface area of an element (of both faces if the FEA data contains them).

- string **TimeStringFromSeconds** (double seconds, double daysinyear)
  Returns a human readable string for the length of a time period given in seconds.

- double **U** (double Y, double K_Ic, double v_0, double n)
  Combined fracture mechanics and crack growth coefficient U.

- double **WeibullInv** (double probability, double shape, double scale)
  Returns the inverse cumulative distribution function for a Weibull distribution (missing in Excel).

- object **WeibullScale** (double mean, double stdev)
  Returns the Weibull scale parameter.

- object **WeibullShape** (double mean, double stdev)
  Returns the Weibull shape parameter.

Member Function Documentation

object MH::GlassFunctionsForExcel::InterExtrapolateLinear (double x, Excel.Range xrange, Excel.Range yrange)

Linear interpolation AND extrapolation using Excel cell ranges.

Parameters:

x   the x value to find y(x) for
xrange the list of x values that define the curve (may be unsorted)
yrange the list of corresponding y values

object MH::GlassFunctionsForExcel::InterpolateLinear (double x, Excel.Range xrange, Excel.Range yrange)

Linear interpolation using Excel cell ranges.

Parameters:

x   the x value to find y(x) for
xrange the list of x values that define the curve (may be unsorted)
yrange the list of corresponding y values

string MH::GlassFunctionsForExcel::TimeStringFromSeconds (double seconds, double daysinyear)

Returns a human readable string for the length of a time period given in seconds.

Parameters:

seconds   the time period in seconds
daysinyear the number of days in a year to use for conversion (usually 356)

H.3 MH::SSFM Class Reference

Collaboration diagram for MH::SSFM:

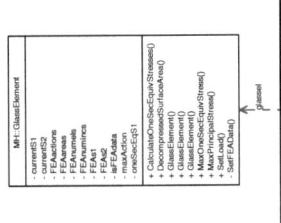
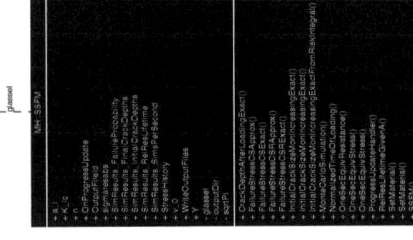

Detailed Description

This class represents the single surface flaw model (**SSFM**(p. 183)).

Public Member Functions

- double **CrackDepthAfterLoadingExact** (double sigma_eq_1s)

  Returns the crack depth after loading for some 1s-equivalent static stress (exact formulation). If the crack fails, -1 is returned.

- double **FailureStressCSApprox** (double timetofailure)

  Approximate formulation of FailureStressCSExact; assuming $a_{final} \gg a_i$ (enables direct solution).

- object **FailureStressCSExact** (double timetofailure)

  Failure stress for constant stress loading (exact formulation) (Calculation using bracketing of InitialCrackSizeExactFormulation).

- double **FailureStressCSRApprox** (double stressrate)

  Approximate formulation of FailureStressCSRExact; assuming $a_{final} \gg a_i$ (enables direct solution).

- double **FailureStressCSRExact** (double stressrate)

  Failure stress for constant stress rate loading (exact formulation) (Calculation using bracketing of InitialCrackSizeExactFormulation).

- double **InitialCrackSizeMonIncreasingExact** (double sigma_eq_-timetofailure, double failurestress, double timetofailure)

  Initial crack size that leads to failure after a MONOTONOUSLY INCREASING (or constant) stress history (exact formulation); Note: alternative parameter set for backwards-compatibility.

- double **InitialCrackSizeMonIncreasingExact** (double sigma_eq_1s, double failurestress)

  Initial crack size that leads to failure after the MONOTONOUSLY INCREASING (or constant) stress history characterized by the one-second equivalent static stress and the failure stress (exact formulation).

- double **InitialCrackSizeMonIncreasingExactFromRiskIntegral** (double riskintvalue, double failurestress)

Public Attributes

- **double a_i**
  Initial crack size.
- **double K_Ic**
  Fracture toughness.
- **double n**
  Exponential crack velocity parameter.
- **ProgressUpdateHandler OnProgressUpdate**
  An instance of the ProgressUpdateHandler delegate.
- **string OutputFileId**
  An id string to add to output filenames. By setting this, the overwriting of output files by successive calculations can be avoided.
- **double sigmaresabs**
  The absolute value of the residual surface stress.
- **double SimResults_FailureProbability**
  The failure probability that resulted from the last simulation run (failures / realizations).
- **double[] SimResults_FinalCrackDepths**
  Final crack depth results from all simulation runs; -1 means failure.
- **double[] SimResults_InitialCrackDepths**
  Initial crack depths from all simulation runs.
- **double[] SimResults_RelResLifetime**
  Relative residual lifetime from all simulation runs.
- **double SimResults_SimsPerSecond**
  Performance indicator: The number of simulation runs per second.
- **double[,] StressHistory**
  Stress history (duration, stress).
- **double v_0**
  Linear crack velocity parameter.
- **bool WriteOutputFiles** = false
  Enables or disables result output to tab separated text files.
- **double Y**
  Geometry factor.

Initial crack size that leads to failure after the MONOTONOUSLY INCREASING (or constant) stress history characterized by the value of the risk integral and the failure stress (exact formulation) Note: risk-integral based parameter set for performance in simulations.

- void **MonteCarloSimulation** (uint realizations, int actHistLengthInYears, **ActionObj** action, **MaterialObj** material, bool exactformulation)
  Perform a Monte Carlo simulation using an action and a material object that may both consist of random variables (probability distributions).
- double **NormalizedTimeOfLoading** (double[,] actionhistory)
  The normalized time of loading.
- double **OneSecEquivResistance** ()
  The one-second equivalent resistance of the single crack.
- double **OneSecEquivStress** (double stress, double duration, double n)
  The one-second equivalent stress for some constant stress.
- double **OneSecEquivStress** (double[,] actionhistory)
  The one-second equivalent stress for a given action history.
- delegate void **ProgressUpdateHandler** (int progressPercent)
  A delegate to indicate progress of calculations (in percent).
- double **RelResLifetimeGivenA** (double a)
  Relative residual lifetime as a function of the momentary crack size.
- void **SetMaterial** (double sigmaresabs, double a_i, double Y, double K_Ic, double v_0, double n)
  Set the material properties using individual parameters.
- void **SetMaterial** (**MaterialObj** material)
  Set the material properties using a **MaterialObj**(p. 190) object.
- **SSFM** (double sigmaresabs, double a_i, double Y, double K_Ic, double v_0, double n)
  Special constructor: individual parameters, action = stress (no GlassElement(p. 187) required).
- **SSFM** (**GlassElement** glassel, **MaterialObj** material)
  Constructor for a single parameter of type ResistanceObject.

H.4 MH::RSFPM Class Reference

Collaboration diagram for MH::RSFPM:

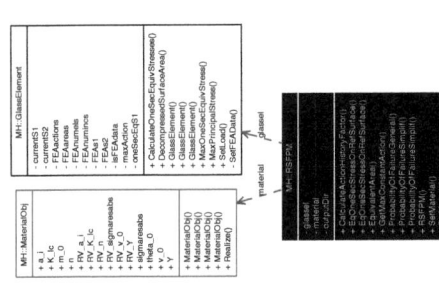

Detailed Description

This class represents the random surface flaw population model (**RSFPM**(p. 185)).

Public Member Functions

- double **CalculateActionHistoryFactor** (double pfAtEndOfLt, double[,] actionHistory)

    Finds the action history factor for a given action history that yields a given failure probability (using the simplified formulation of the failure probabilty).

Properties

- double **LastSafetyIndex**

    The safety index (beta value) calculated from SimResults_failureProbability (assumes a normal distribution).

- string **OutputDirectory**

    The directory for file output.

Member Function Documentation

void MH::SSFM::MonteCarloSimulation (uint realizations, int actHistLengthInYears, ActionObj action, MaterialObj material, bool exactformulation)

Perform a Monte Carlo simulation using an action and a material object that may both consist of random variables (probability distributions).

Parameters:

realizations the number of simulation runs to perform

actHistLengthInYears the duration of the action histories that are generated

action the action object

material the material object

exactformulation true: use the exact formulation; false: use the simplified formulation assuming $a_{final} \gg a_i$

- double **EqOneSecStressOnRefSurface** (double actionDuration, double actionIntensity)

    Returns the equivalent one-second uniform first principal stress on the reference surface ($\bar{\sigma}$) for a constant action.

- double **EqOneSecStressOnRefSurface** (double[,] actionHistory)

    Returns the equivalent one-second uniform first principal stress on the reference surface ($\bar{\sigma}$) for any action history.

- double **EquivalentArea** (double actionIntensity)

    Returns the equivalent surface area $\bar{A}$ as a function of the action intensity.

- double **GetMaxConstantAction** (int reqLt, double pfAtEndOfLt)

    Finds the constant action of a given duration that yields a given failure probability.

- double **ProbabilityOfFailureGeneral** (double[,] actionHistory, int numOfSegments)

    Returns the probability of failure (general formulation).

- double **ProbabilityOfFailureSimplif** (double[,] actionHistory, double actHistFactor)

    Returns the probability of failure (simplified formulation using $\bar{\sigma}$); Overload taking an action history and an action history factor as arguments.

- double **ProbabilityOfFailureSimplif** (double sigma_bar)

    Returns the probability of failure (simplified formulation using $\bar{\sigma}$).

- **RSFPM** (GlassElement glassel, **MaterialObj** material)

    This class represents the random surface flaw population model.

- void **SetMaterial** (double sigmaresabs, double theta_0, double m_0, double Y, double K_Ic, double v_0, double n)

    Sets the material properties from individual parameters.

- void **SetMaterial** (**MaterialObj** material)

    *Sets the material properties from a **MaterialObj**(p. 190) object.*

Properties

- string **OutputDirectory**

    The directory for file output.

Member Function Documentation

double MH::RSFPM::CalculateActionHistoryFactor (double pfAtEndOfLt, double actionHistory[,])

Finds the action history factor for a given action history that yields a given failure probability (using the simplified formulation of the failure probability).

Parameters:

  pfAtEndOfLt failure probabilitiy at the end of the action history
  actionHistory the action history

double MH::RSFPM::GetMaxConstantAction (int reqLt, double pfAtEndOfLt)

Finds the constant action of a given duration that yields a given failure probability.

Parameters:

  reqLt the required lifetime (s)
  pfAtEndOfLt the probability of failure at the end of the lifetime

double MH::RSFPM::ProbabilityOfFailureGeneral (double actionHistory[,], int numOfSegments)

Returns the probability of failure (general formulation).

Parameters:

  actionHistory the action history
  numOfSegments the number of segments to use for biaxial stress field calculation

H.5 MH::GlassElement Class Reference

Detailed Description

This class represents a glass element. It requires in general finite element data representing the action/stress relationship. The main purpose of the class is to calculate the stresses at all points on the element surface as a function of the action.

Public Member Functions

- void **CalculateOneSecEquivStresses** (double[,] ActionHistory, double sigmaresabs, double n)

  Calculate the one-second equivalent stresses for all elements, $\sigma_{1,1s,i}$. The results are available through the this.OneSecEqS1 property.

- double **DecompressedSurfaceArea** ()

  The decompressed surface area for a given load level taking residual stress into account (max. of all finite elements). IMPORTANT: Set the load magnitude by SetLoad before calling this!

- **GlassElement** (string fea1path, string fea2path)

  Constructor for a glass element; takes paths to FEA data_files as arguments.

- **GlassElement** (double[,] feadata1, double[,] feadata2)

  Constructor for a glass element; takes arrays as arguments.

- **GlassElement** ()

  Special constructor: creates a 'dummy' glass element with one finite element and without FEA data; Can be used for the SSFM(p. 183) if the action intensity is given in terms of the max. in-plane principal stress (σ_1).

- double **MaxOneSecEquivStress** (double[,] ActionHistory, double sigmaresabs, double n)

  Get the maximum one-second equivalent stress, max($\sigma_{.,1s,i}$).

- double **MaxPrincipalStress** ()

  The max. in-plane principal stress due to external load AND residual stress (max. of all finite elements). IMPORTANT: Set the load magnitude by SetLoad before calling this!

- void **SetLoad** (double actionMagnitude, double sigmaresabs, bool withsigma2)

  Calculate the effective crack opening surface stress in all elements for a given action intensity.

Properties

- double[] **CurrentS1**

  The current max. in-plane principal stress values (incl. residual stress).

- double[] **CurrentS2**

  The current min. in-plane principal stress values (incl. residual stress).

- double[] **FEAAreas**

  The areas of all finite elements.

- double **MaxAction**

  The maximum action that there is FEA data for.

- int **NumOfElements**

  The number of finite elements in the FEA data.

- int **NumOfIncs**

  The number of increments in the FEA data.

- double[] **OneSecEqS1**

  The current 1s-equivalent first in-plane principal stress values.

- double **TotalSurfaceArea**

  Total surface area of the element.

Constructor & Destructor Documentation

MH::GlassElement::GlassElement (double *feadata1*[,], double *feadata2*[,])

Constructor for a glass element; takes arrays as arguments.

Parameters:

feadata1 Array with the FEA data for σ_1
feadata2 Array with the FEA data for σ_2 if it shall be considered, null if not.

H.6 MH::ActionObj Class Reference

Collaboration diagram for MH::ActionObj:

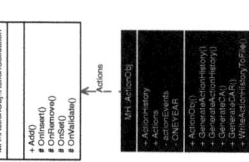

Detailed Description

This class represents a probabilistic or deterministic action model and generates action histories from this model. Random variables are represented by instances of the probability distribution classes.

Public Member Functions

- **ActionObj** ()

  Constructor.

- void **GenerateActionHistory** (int *lengthinyears*, double *actionfactor*)

  Generates an action history of a given length using the action definitions in this.Actions and an action factor.

- void **GenerateActionHistory** (int *lengthinyears*)

  Generates an action history of a given length using the action definitions in this.Actions.

- void **GenerateCA** (double *duration*, double *intensity*)

  Generate an ActionHistory consisting of one constant action.

MH::GlassElement::GlassElement (string *fea1path***, string** *fea2path***)**

Constructor for a glass element; takes paths to FEA data files as arguments.

Parameters:

fea1path path of file containing the FEA data for σ_1

fea2path path of file containing the FEA data for σ_2; empty string if only sigma_1 shall be considered

Member Function Documentation

void MH::GlassElement::SetLoad (double *actionMagnitude***, double** *sigmaresabs***, bool** *withsigma2***)**

Calculate the effective crack opening surface stress in all elements for a given action intensity.

Parameters:

actionMagnitude the action intensity

sigmaresabs the absolute value of the residual surface compression stress

withsigma2 true if the min. in-plane principal stress (σ_2) should be considered, false if not

H.7 MH::ActionObj::Action Class Reference

Collaboration diagram for MH::ActionObj::Action:

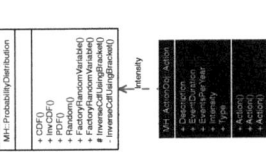

Detailed Description

Each instance of this class represents an action (random variable). Instances are read-only.

- void **GenerateCAR** (double maxaction, double actionrate, int numofsteps)

  Generates an action history for a constant action rate.

- void **WriteActionHistoryToFile** (string filePath)

  Writes the action history to a tab separated file.

Public Attributes

- double[,] **ActionHistory**

  The action history (duration, intensity).

- ActionCollection **Actions** = new ActionCollection()

  The collection of actions.

H.8 MH::ActionObj::ActionCollection Class Reference

Detailed Description

Strongly typed collection for actions.

Classes

- class **Action**

  Each instance of this class represents an action (random variable). Instances are read-only.

- class **ActionCollection**

  Strongly typed collection for actions.

H.9 MH::MaterialObj Class Reference

Collaboration diagram for MH::MaterialObj:

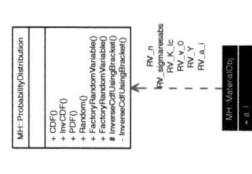
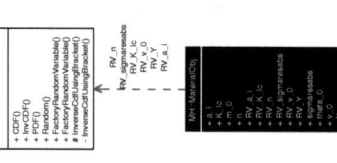

Detailed Description

This class represents a probabilistic or deterministic material model. Random variables are represented by instances of the probability distribution classes (call **Realize()**(p. 190) to realize them all).

Public Member Functions

- **MaterialObj** (double **sigmaresabs**, double **m_0**, double **n**)

  Constructor for a deterministic material object with a reduced parameter set.

- **MaterialObj** (double **sigmaresabs**, double **theta_0**, double **m_0**, double **Y**, double **K_Ic**, double **v_0**, double **n**)

  Constructor for a deterministic material object to be used with the RSFPM(p. 185).

- **MaterialObj** (double **sigmaresabs**, double **a_i**, double **Y**, double **K_Ic**, double **v_0**, double **n**)

  Constructor for a deterministic material object to be used with the SSFM(p. 183).

- **MaterialObj** ()

  Constructor for an empty material object.

- void **Realize** ()

  Fill all properties of this class with realization of the corresponding random variables (if they exist).

Public Attributes

- double **a_i**

  Initial crack depth.

- double **K_Ic**

  Fracture toughness.

- double **m_0**

  Surface condition parameter m_0.

- double **n**

  Exponential crack velocity parameter.

- **ProbabilityDistribution RV_a_i**

  Random variable of the initial crack depth.

- **ProbabilityDistribution RV_K_Ic**

  Random variable of the fracture toughness.

- **ProbabilityDistribution RV_n**

  Random variable of the exponential crack velocity parameter.

- **ProbabilityDistribution RV_sigmaresabs**

  Random variable of the absolute value of the residual surface stress.

- **ProbabilityDistribution RV_v_0**

  Random variable of the linear crack velocity parameter.

H.10 MH::ProbabilityDistribution Class Reference

Inheritance diagram for MH::ProbabilityDistribution:

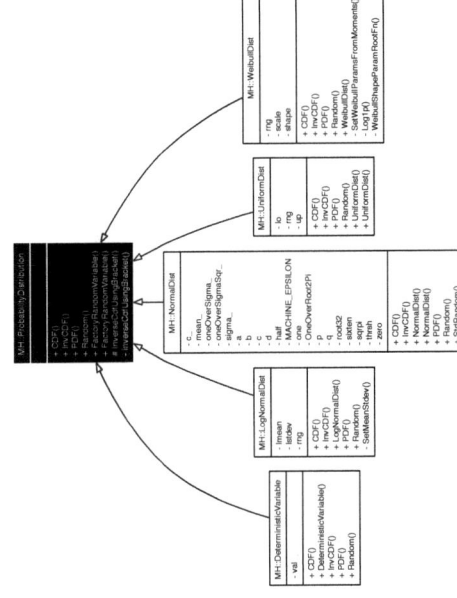

Detailed Description

Abstract class for all probability distribution classes.

Public Types

- enum **ProbDistType**

  Enumeration containing the available probability distribution types.

- **ProbabilityDistribution RV_Y**

  Random variable of the geometry factor.

- double **sigmaresabs**

  Absolute value of the residual surface stress.

- double **theta_0**

  Surface condition parameter θ_0.

- double **v_0**

  Linear crack velocity parameter.

- double **Y**

  Geometry factor.

Properties

- double **f**

  The f geometry factor defined as $f = Y \cdot \sqrt{\pi}$.

- double **K_bar**

  The derived surface flaw parameter $\bar{k} = f(U, \theta_0, m_0, n)$.

- double **M_bar**

  The surface flaw parameter $\bar{m} = f(m_0, n)$.

- double **S**

  The combined crack velocity parameter $S = v_0 \cdot K_{Ic}^{-n}$.

- double **U**

  The crack growth related constant $U = f(K_{Ic}, n, v_0, Y)$.

Public Member Functions

- abstract double **CDF** (double x)

  The cumulative distribution function.

- abstract double **InvCDF** (double p)

  The inverse cumulative distribution function.

- abstract double **PDF** (double x)

  The probability density function.

- abstract double **Random** ()

  Sample a random value from the probability distribution.

Static Public Member Functions

- static **ProbabilityDistribution FactoryRandomVariable** (string disttype, double mean, double cov)

  Factory a random variable.

- static **ProbabilityDistribution FactoryRandomVariable** (RVParameterSet ps)

  Factory a random variable.

Protected Member Functions

- double **InverseCdfUsingBracket** (double p, double lowerBound, double upperBound)

  Find the inverse of a cumulative distribution function using a bracketing algorithm.

Properties

- double **CharacteristicValue**

  The characteristic value (set CharValProb first!).

- double **CharValProb**

  The cumulative probability of the characteristic value.

- abstract double **Mean**

  Mean.

- abstract double **Stdev**

  Standard deviation.

- abstract double **Variance**

  Variance.

Classes

- class **RVParameterSet**

  The parameter set for a random variable.

Member Function Documentation

static ProbabilityDistribution MH::ProbabilityDistribution::FactoryRandomVariable (**string** *disttype*, **double** *mean*, **double** *cov*) [static]

Factory a random variable.

Parameters:

disttype	the distribution type (Deterministic, Uniform, Normal, Lognormal or Weibull)
mean	the mean value
cov	the coefficient of variation

double MH::ProbabilityDistribution::InverseCdfUsingBracket (**double** *p*, **double** *lowerBound*, **double** *upperBound*) [protected]

Find the inverse of a cumulative distribution function using a bracketing algorithm.

Parameters:

p	probability
lowerBound	lower bound for the inverse value
upperBound	upper bound for the inverse value

Returns:

The value x such that CDF(x) = p.

H.11 MH::NormalDist Class Reference

Detailed Description

Normal (Gaussian) probability distribution.

Public Member Functions

- override double **CDF** (double x)
  The cumulative distribution function.
- override double **InvCDF** (double p)
  The inverse cumulative distribution function.
- **NormalDist** ()
  Constructs a standard normal distribution (mean = 0, variance = standard deviation = 1).
- **NormalDist** (double mean, double stdev)
  Constructs a normal distribution (Gaussian distribution) instance with the given mean and standard deviation.
- override double **PDF** (double x)
  The probability density function.
- override double **Random** ()
  Random sample of the normal distribution (with E(X)=Mean ; Stdev(X)=Stdev).
- double **StdRandom** ()
  Random sample of the standard normal distribution.

Properties

- double **Kurtosis**
  Gets the kurtosis, a measure of the degree of peakedness of the density.
- override double **Mean**
  Gets and sets the mean of the density.
- double **Skewness**
  Gets the skewness, a measure of the degree of asymmetry of this density.
- override double **Stdev**
  Gets and sets the standard deviation of the density.
- override double **Variance**
  Gets and sets the variance of the density.

H.12 MH::LogNormalDist Class Reference

Detailed Description

Log-normal probability distribution.

Public Member Functions

- override double **CDF** (double x)
  Returns the cumulative density function evaluated at a given value.
- override double **InvCDF** (double p)
  The inverse cumulative distribution function.
- **LogNormalDist** (double mean, double stdev)
  Constructs a log-normal distribution instance with the given mean and standard deviation.
- override double **PDF** (double x)
  The probability density function.
- override double **Random** ()
  Sample a random value from the probability distribution.

Properties

- double **LMean**
  Gets or sets the mean of the distribtion of ln(x).
- double **LStdev**
  Gets or sets the standard deviation of the distribition of ln(x).
- override double **Mean**

Mean.

- override double **Stdev**
  Standard deviation.
- override double **Variance**
  Variance.

Constructor & Destructor Documentation

MH::LogNormalDist::LogNormalDist (double *mean*, double *stdev*)

Constructs a log-normal distribution instance with the given mean and standard deviation.

Parameters:
mean The mean of x (NOT log(x)).
stdev The standard deviation of x (NOT log(x)).

H.13   MH::UniformDist Class Reference

Detailed Description

Uniform (flat) probability distribution over [lower_boud, upper_bound].

Public Member Functions

- override double **CDF** (double x)
  The cumulative distribution function.
- override double **InvCDF** (double p)
  The inverse cumulative distribution function.
- override double **PDF** (double x)
  The probability density function.
- override double **Random** ()
  Sample a random value from the probability distribution.
- **UniformDist** (double lower_bound, double upper_bound)
  Constructor for a uniform distribution over [lower_bound, upper_bound].
- **UniformDist** ()
  Constructor for a standard uniform distribution [0,1].

Properties

- override double **Mean**
  Mean.
- override double **Stdev**
  Standard deviation.
- override double **Variance**
  Variance.

H.14   MH::WeibullDist Class Reference

Detailed Description

Weibull probability distribution.

Public Member Functions

- override double **CDF** (double x)
  The cumulative distribution function.
- override double **InvCDF** (double p)
  The inverse cumulative distribution function.
- override double **PDF** (double x)
  The probability density function.
- override double **Random** ()
  Sample a random value from the probability distribution.
- **WeibullDist** (double mean, double stdev)
  Constructs a weibull distribution instance with the given mean and standard deviation.

Properties

- override double **Mean**
  Mean.
- double **Scale**
  The Weibull scale parameter.
- double **Shape**
  The Weibull shape parameter.
- override double **Stdev**
  Standard deviation.
- override double **Variance**
  Variance.

H.15 MH::DeterministicVariable Class Reference

Detailed Description

A deterministic variable (a single scalar value) implementing the **ProbabilityDistribution**(p. 191) abstract class. Instances can be used to supply deterministic values to functions that expect a **ProbabilityDistribution**(p. 191).

Public Member Functions

- override double **CDF** (double x)
  The cumulative distribution function.
- **DeterministicVariable** (double Value)
  Constructor.
- override double **InvCDF** (double p)
  InvCDF can only be evaluated for the probability $p = 1$.
- override double **PDF** (double x)
  The probability density function.
- override double **Random** ()
  Always returns the value of the variable (deterministic).

Properties

- override double **Mean**
  Mean.
- override double **Stdev**
  Standard deviation.
- double **Value**
  The value of the deterministic variable.
- override double **Variance**
  Variance.

Index

4PB, xiv

Abaqus, 90
action, xv
action history, xv
action history effect, 30
action history generator, 116
action intensity, xv
action intensity history, 116
action model, 116, 133
addend, xv
aging, 40
air side, xv
air side (of glass), 6
alkali, xv
alkali leaching, 41
allowable stress, 17
ambient strength, 58
ambient strength data, 86
ambient temperature, xv
Anderson-Darling test, 58, 168
ANG, xiv
annealed glass, 10
annealing, xv, 6
artificially induced surface damage, xv
as-received glass, xv
aspect ratio, xv
ASTM E 1300, 25
autoclave, xv
average refractive index, 8

beta distribution, 166
bevelling, xv
biaxial stress correction factor, 25, 33, 67
biaxial stress field, 52, 66, 129
birefringence, 103
blast-resistant glass, 14
borosilicate glass, 6

breakage stress, 16
Brown's integral, *see* risk integral
BSG, xiv
bullet-resistant glass, 14

C++, 131
C#, 131
CAN/CGSB 12.20-M89, 27
CDF, xiv
CDR, xiv
characteristic crack propagation speed, 39
chemical composition, 6, 42
chemical reaction at the crack tip, 40
close-to-reality surface flaws, 105
coating, xv
coaxial double ring test, 15, 16, 66, 75, 98
coefficient of thermal expansion, 8
coefficient of variation (CoV), xv
computing time, xv
concentric ring-on-ring test, *see* coaxial double ring test
confocal microscope, 103
constant load rate loading, xv
constant load rate testing, 15
constant stress rate loading, xv
constant stress rate testing, 15, 75
constant stress testing, 75
contact damage, 108
convection, xvi
corrosive media, 41
crack, xvi, 43
crack depth, 43
crack depth at failure, 46
crack front, 43
crack growth limit, *see* crack growth threshold
crack growth threshold, 40, 64, 80

crack healing, 32, 40
crack length, 43
crack opening stress, 45, 72, 102
crack orientation, 53
crack repropagation, 40
crack tip, 43
crack tip blunting, 41
crack velocity, 38
crack velocity parameters, 82
critical crack depth, 44, 49
critical stress, 44
critical stress intensity factor, *see* fracture toughness
CS, xiv
CSR, xiv
cullet, xvi
curtain walling, xvi

damage accumulation, 62
DCB, xiv
decompressed surface, xvi, 32
deep surface flaws, 108, 126
defect, xvi
deflection, xvi
deformation, xvi
DELR design method, 18
density, 8
desiccants, xvi
design, *see* structural design
design flaw, 117, 123, 126
design life, xvi
design method of damage equivalent load and resistance, *see* DELR design method
design point, 119
design process, 116
design resistance, 119
detectability, *see* visual detectability

differential stress refractometer, 103
dimensionless stress distribution function, 72
direct crack growth measurement, 17
double glazing, double-glazed units, xvi
DT, xiv
duration-of-load effect, *see* load duration effect
dynamic fatigue test, 15
dynamic viscosity, 8

effective area, *see* equivalent area
effective nominal flaw depth, xvi, 103, 108, 110
elasticity, xvi
elongation, xvi
emissivity, xvi, 8
empirical cumulative distribution function, 166
empirical probability of failure, 166
energy release rate, 44
environmental fatigue, 37
equibiaxial stress field, xvi, 16, 33, 61, 66
equivalent t_0-second uniform stress on the unit surface area, 71
equivalent area, 73
equivalent reference stress, 71, 72, 74
equivalent representative stress, 73, 74
equivalent resistance, 47
equivalent stress, 46
equivalent stress intensity factor, 52
equivalent uniformly distributed stress, 71
estimator, 166
European design methods, 29, 76, 119
expectation value, 166
exponential crack velocity parameter, 39
exposed surface, 123

face, xvi
failure probability, 49
failure stress, 16
fatigue limit, *see* crack growth threshold
FE, xiv
fire protection glass, 9, 14
flat glass, xvi
flaw, xvi

flaw location, 61
flaw orientation, 61
float glass, xvi
float process, 5
four point bending test, 15, 16, 75
fracture mode, 52
fracture pattern, 10
fracture toughness, 38, 43, 80
fractured glass zone, 110
FTG, xiv
fully tempered glass, xvi, 10, 11, 45
furnace, 6

geometry factor, 43, 80
GFPM, xiv, *see* glass failure prediction model
glass, xvi
glass failure prediction model, 25
glass pane, 9
glass products, 9
glass type, 10
glass type factor, 26
glass unit, 9
GlassFunctionsForExcel, 132, 169
GlassTools, 128
glazing, xvi
goodness-of-fit, 58
greenhouse effect, 8
Griffith, 37, 43

half-penny shaped crack, 81
hardness, xvi
heat strengthened glass, xvii, 10, 12, 45
heat treated glass, xvii, 45
heat-soak test (HST), xvi
hermetic coating, 104, 110, 126
homogeneous, xvii
HSG, xiv
humidity, 41
hysteresis effect, 40

impact, xvii
impact strength, xvii
in-plane principal stress, *see* principal stress
indentation crack, 16
inert conditions, 104
inert failure probability, 51
inert fatigue, 37
inert strength, 44, 58, 89, 102
inert testing, 76, 104
inherent strength, xvii, 21, 30, 32
initial crack depth, 45
initial surface condition, 62
inspection, 124
insulating glass unit, 9
insulating glass unit (IGU), xvii

interlayer, xvii
intumescence, xvii
IPP, xiv
Irwin, 43
Irwin's fracture criterion, 43

joint, xvii

Knoop hardness, 8
Kolmogorov-Smirnov test, 168

laminated glass, xvii, 9
large through-thickness crack, 82
lateral load, xvii
LDSR, *see* differential stress refractometer
least squares method, 90, 167
LEFM, xiv
lehr, xvii, 6
Levengood, 37
lifetime, 46
lifetime prediction, xvii
lifetime prediction model, 116
linear crack velocity parameter, 39
linear elastic fracture mechanics, 43
load bearing capacity, 116
load duration effect, 30, 31
load duration factor, xvii
load shape, xvii
loading ring, 98
loading time, xvii
log-normal distribution, 58, 165
long, straight-fronted plane edge crack, 81
long-term loading, 123
low emissivity coating (low-e coating), xvii
low iron glass, xvii
lower visual detectability limit, 112
LR, xiv

machining damage, 61
material model, 116
Matlab, 90, 131
maximum likelihood method, 90, 167
mean rank, 166
median, 166
median rank, 166
melting temperature, 7
method of moments, 167
mode I, xvii
model hypotheses, 57
momentary critical crack depth, 54
monotonously increasing, xvii
multimodal failure criterion, 52
multiple-glazed units, xvii

near-inert conditions, 46, 104
nickel sulfide inclusion, xvii
non-exposed surfaces, 124
non-factored load, 26
non-uniform stress field, xvii, 52
normal distribution, 58, 119, 165
normalized lifetime, 47
normalized loading time, 47
normalized pressure, 68
normalized residual lifetime, 48
North American design methods, 29, 76, 119

observed significance level probability, 168
Ontario Research Foundation, 88
optical flaw depth, xvii, 103, 108, 110
optical properties, 8
optical quality, 13
ORF, xiv

p-value, 168
pane (of glass), xvii
Pareto distribution, 50, 60, 165
PDF, xiv
performance requirements, 116
pH value, 41
physical properties, 8
point estimate, 167
Poisson distribution, 50, 166
Poisson's ratio, 8
polyvinyl butyral, see PVB
predictive modelling, xvii, 116
prEN 13474, 20
principal stress, 53
principal stress ratio, 67, 68
profile glass, xvii
PVB, xiv, xviii
Python, 131

quality control, 124
quarter circle crack, 82

R400 test setup, 16, 98
R45 test setup, 98
radiation, xviii
random surface flaw population, 49, 60, 116, 118
random variable, 166
reaction ring, 98
realization, xviii
reference ambient strength, 72
reference inert strength, 51, 119
reference time period, 46, 77
reflection, 8
relative reduction of lifetime, 48
relative residual lifetime, 48

reliability index, 119
renucleation, 41
representative stress, 72
residual stress, xviii, 32
residual surface stress, 45, 92
rigidity, xviii
risk integral, 31, 46, 62
RSFP, xiv, see random surface flaw population

safety glass, 10
sandblasting, xviii
sandpaper scratching, 82
scale parameter, 17, 51
SCG, xiv
self-fatigue, 12
severe damage, 123
shape parameter, 17, 51
Shapiro-Wilk test, 168
Shen, 22
short-term loading, 124
SIF, xiv
silica, xviii
silicates, xviii
single surface flaw, 116, 117, 123
size effect, 32, 51
slow crack growth, 37
SLS, xiv
soda lime silica glass, 6
solidification, 7
spacer, spacer bar, xviii
specific thermal capacity, 8
SSF, xiv, see single surface flaw
static fatigue, 37
static fatigue test, 15
strength, xviii
stress corrosion, 37
stress corrosion limit, see crack growth threshold
stress distribution function, see dimensionless stress distribution function
stress intensity factor, 38, 43
stress rate, xviii
structural design, xviii, 71, 116
structural glazing, xviii
structural model, 116
structural sealant glazing, xviii
subcritical crack growth, 37, 54, 115
supply rate, 40
surface condition, 115
surface condition parameters, 125
surface crack, 43
surface damage hazard scenario, 115
surface decompression, 45, 115

surface scratching device, 106
survival probability, 50

target failure probability, 51, 71, 72, 116, 119
target reliability index, 119
temperature, 41
tempering, 10
    chemical, 12
    thermal, 11
tensile strength, xviii, 8
tensile strength ratio, 74
test setup, 98
testing, 125
thermal conductivity, 8
thermal expansion coefficient, 7
thermal stress, xviii
threshold crack depth, 65
threshold stress intensity, see crack growth threshold
threshold stress intensity factor, 65
through-thickness crack, 17
time to failure, see lifetime
time-dependent failure probability, 55
time-dependent loading, 54
tin bath, 6
tin side, xviii, 6, 103
tinted glass, 7
transformation temperature, 7
transient analysis, xviii
transient finite element analysis, 72
transparency, 8
transparent, xviii
TRAV, 18
TRLV, 18

uniaxial stress field, xviii, 16, 34, 66
uniform distribution, 53, 60, 165
uniform lateral load, xviii
uniform stress field, xviii
unit surface area, 51

vacuum, 104
viscosity, xviii, 7
visual detectability, 111
visual inspection, 111
volume crack, 43

Weibull distribution, 16, 51, 58, 165
Weibull parameters, 167
Weibull probability plot, 58

Young's modulus, 8

Die VDM Verlagsservicegesellschaft sucht für wissenschaftliche Verlage abgeschlossene und herausragende

Dissertationen, Habilitationen, Diplomarbeiten, Master Theses, Magisterarbeiten usw.

für die kostenlose Publikation als Fachbuch.

Sie verfügen über eine Arbeit, die hohen inhaltlichen und formalen Ansprüchen genügt, und haben Interesse an einer honorarvergüteten Publikation?

Dann senden Sie bitte erste Informationen über sich und Ihre Arbeit per Email an *info@vdm-vsg.de*.

Sie erhalten kurzfristig unser Feedback!

VDM Verlagsservicegesellschaft mbH
Dudweiler Landstr. 99   Telefon +49 681 3720 174
D - 66123 Saarbrücken   Fax   +49 681 3720 1749
www.vdm-vsg.de

Die VDM Verlagsservicegesellschaft mbH vertritt

Printed by Books on Demand GmbH, Norderstedt / Germany